Multiple Imputation for
Nonresponse in Surveys

Probability and Mathematical Statistics (Continued)
PURI, VILAPLANA, and WERTZ • New Perspectives in Theoretical and Applied Statistics
RANDLES and WOLFE • Introduction to the Theory of Nonparametric Statistics
RAO • Linear Statistical Inference and Its Applications, *Second Edition*
RAO • Real and Stochastic Analysis
RAO and SEDRANSK • W.G. Cochran's Impact on Statistics
RAO • Asymptotic Theory of Statistical Inference
ROHATGI • An Introduction to Probability Theory and Mathematical Statistics
ROHATGI • Statistical Inference
ROSS • Stochastic Processes
RUBINSTEIN • Simulation and The Monte Carlo Method
SCHEFFE • The Analysis of Variance
SEBER • Linear Regression Analysis
SEBER • Multivariate Observations
SEN • Sequential Nonparametrics: Invariance Principles and Statistical Inference
SERFLING • Approximation Theorems of Mathematical Statistics
SHORACK and WELLNER • Empirical Processes with Applications to Statistics
TJUR • Probability Based on Radon Measures

Applied Probability and Statistics
ABRAHAM and LEDOLTER • Statistical Methods for Forecasting
AGRESTI • Analysis of Ordinal Categorical Data
AICKIN • Linear Statistical Analysis of Discrete Data
ANDERSON, AUQUIER, HAUCK, OAKES, VANDAELE, and WEISBERG • Statistical Methods for Comparative Studies
ARTHANARI and DODGE • Mathematical Programming in Statistics
ASMUSSEN • Applied Probability and Queues
BAILEY • The Elements of Stochastic Processes with Applications to the Natural Sciences
BAILEY • Mathematics, Statistics and Systems for Health
BARNETT • Interpreting Multivariate Data
BARNETT and LEWIS • Outliers in Statistical Data, *Second Edition*
BARTHOLOMEW • Stochastic Models for Social Processes, *Third Edition*
BARTHOLOMEW and FORBES • Statistical Techniques for Manpower Planning
BECK and ARNOLD • Parameter Estimation in Engineering and Science
BELSLEY, KUH, and WELSCH • Regression Diagnostics: Identifying Influential Data and Sources of Collinearity
BHAT • Elements of Applied Stochastic Processes, *Second Edition*
BLOOMFIELD • Fourier Analysis of Time Series: An Introduction
BOX • R. A. Fisher, The Life of a Scientist
BOX and DRAPER • Empirical Model-Building and Response Surfaces
BOX and DRAPER • Evolutionary Operation: A Statistical Method for Process Improvement
BOX, HUNTER, and HUNTER • Statistics for Experimenters: An Introduction to Design, Data Analysis, and Model Building
BROWN and HOLLANDER • Statistics: A Biomedical Introduction
BUNKE and BUNKE • Statistical Inference in Linear Models, Volume I
CHAMBERS • Computational Methods for Data Analysis
CHATTERJEE and PRICE • Regression Analysis by Example
CHOW • Econometric Analysis by Control Methods
CLARKE and DISNEY • Probability and Random Processes: A First Course with Applications, *Second Edition*
COCHRAN • Sampling Techniques, *Third Edition*
COCHRAN and COX • Experimental Designs, *Second Edition*

Applied Probability and Statistics (Continued)

CONOVER • Practical Nonparametric Statistics, *Second Edition*
CONOVER and IMAN • Introduction to Modern Business Statistics
CORNELL • Experiments with Mixtures: Designs, Models and The Analysis of Mixture Data
COX • Planning of Experiments
COX • A Handbook of Introductory Statistical Methods
DANIEL • Biostatistics: A Foundation for Analysis in the Health Sciences, *Fourth Edition*
DANIEL • Applications of Statistics to Industrial Experimentation
DANIEL and WOOD • Fitting Equations to Data: Computer Analysis of Multifactor Data, *Second Edition*
DAVID • Order Statistics, *Second Edition*
DAVISON • Multidimensional Scaling
DEGROOT, FIENBERG and KADANE • Statistics and the Law
DEMING • Sample Design in Business Research
DILLON and GOLDSTEIN • Multivariate Analysis: Methods and Applications
DODGE • Analysis of Experiments with Missing Data
DODGE and ROMIG • Sampling Inspection Tables, *Second Edition*
DOWDY and WEARDEN • Statistics for Research
DRAPER and SMITH • Applied Regression Analysis, *Second Edition*
DUNN • Basic Statistics: A Primer for the Biomedical Sciences, *Second Edition*
DUNN and CLARK • Applied Statistics: Analysis of Variance and Regression, *Second Edition*
ELANDT-JOHNSON and JOHNSON • Survival Models and Data Analysis
FLEISS • Statistical Methods for Rates and Proportions, *Second Edition*
FLEISS • The Design and Analysis of Clinical Experiments
FOX • Linear Statistical Models and Related Methods
FRANKEN, KÖNIG, ARNDT, and SCHMIDT • Queues and Point Processes
GALLANT • Nonlinear Statistical Models
GIBBONS, OLKIN, and SOBEL • Selecting and Ordering Populations: A New Statistical Methodology
GNANADESIKAN • Methods for Statistical Data Analysis of Multivariate Observations
GREENBERG and WEBSTER • Advanced Econometrics: A Bridge to the Literature
GROSS and HARRIS • Fundamentals of Queueing Theory, *Second Edition*
GUPTA and PANCHAPAKESAN • Multiple Decision Procedures: Theory and Methodology of Selecting and Ranking Populations
GUTTMAN, WILKS, and HUNTER • Introductory Engineering Statistics, *Third Edition*
HAHN and SHAPIRO • Statistical Models in Engineering
HALD • Statistical Tables and Formulas
HALD • Statistical Theory with Engineering Applications
HAND • Discrimination and Classification
HOAGLIN, MOSTELLER and TUKEY • Exploring Data Tables, Trends and Shapes
HOAGLIN, MOSTELLER, and TUKEY • Understanding Robust and Exploratory Data Analysis
HOEL • Elementary Statistics, *Fourth Edition*
HOEL and JESSEN • Basic Statistics for Business and Economics, *Third Edition*
HOGG and KLUGMAN • Loss Distributions
HOLLANDER and WOLFE • Nonparametric Statistical Methods
IMAN and CONOVER • Modern Business Statistics
JAGERS • Branching Processes with Biological Applications
JESSEN • Statistical Survey Techniques
JOHNSON • Multivariate Statistical Simulation

(*continued on back*)

Multiple Imputation for Nonresponse in Surveys

DONALD B. RUBIN

Department of Statistics
Harvard University

JOHN WILEY & SONS
New York • Chichester • Brisbane • Toronto • Singapore

Copyright © 1987 by John Wiley & Sons, Inc.

All rights reserved. Published simultaneously in Canada.

Reproduction or translation of any part of this work beyond that permitted by Section 107 or 108 of the 1976 United States Copyright Act without the permission of the copyright owner is unlawful. Requests for permission or further information should be addressed to the Permissions Department, John Wiley & Sons, Inc.

Library of Congress Cataloging in Publication Data:

Rubin, Donald B.
  Multiple imputation for nonresponse in surveys.

  (Wiley series in probability and mathematical statistics. Applied probability and statistics, ISSN 0271-6232)
    Bibliography: p.
    Includes index.
    1. Multiple imputation (Statistics) I. Title.
II. Series.
HH31.2.R83   1987   001.4'225   86-28935
ISBN 0-471-08705-X

Printed in the United States of America

10 9 8 7 6 5 4 3 2 1

To the family of my childhood
and the family of my parenthood

Preface

Multiple imputation is a statistical technique designed to take advantage of the flexibility in modern computing to handle missing data. With it, each missing value is replaced by two or more imputed values in order to represent the uncertainty about which value to impute. The ideas for multiple imputation first arose in the early 1970s when I was working on a problem of survey nonresponse at Educational Testing Service, here summarized as Example 1.1. This work was published several years later as Rubin (1977a).

The real impetus for multiple imputation, however, came from work encouraged and supported by Fritz Scheuren, then of the United States Social Security Administration and now head of the Statistics of Income Division at the United States Internal Revenue Service. His concern for problems of nonresponse in the Current Population Survey led to a working paper for the Social Security Administration (Rubin, 1977b), which explicitly proposed multiple imputation. Fritz's continued support and encouragement for the idea of multiple imputation resulted in (1) an American Statistical Association invited address on multiple imputation (Rubin, 1978a); (2) continued research, such as published in Rubin (1979a); (3) joint work with Fritz and Thomas N. Herzog in the late 1970s, summarized in several papers including Herzog and Rubin (1983); and (4) application of the ideas in 1980 to file matching, which eventually was published as Rubin (1986).

Another important contributor to the development of multiple imputation has been the United States Census Bureau, which several years ago supported the production of a monograph on multiple imputation (Rubin, 1980a). This monograph was the first of four nearly complete drafts that were supposed to become this book.

The second such draft was composed of the collection of chapters distributed to my class on survey nonresponse at the University of Chicago, Winter Quarter 1983. These stopped short of becoming the book primarily because of two Ph.D. students there, Kim Hung Li and Nathaniel Schenker, both of whom wrote theses on aspects of multiple imputation (Li, 1985; Schenker, 1985). Our efforts provided the foundation for the next level of sophistication, and I am extremely grateful for their involvement and for the outstandingly colleageal atmosphere at the University of Chicago, which made this period so productive.

The third draft owed its demise to continued work involving Schenker and two Ph.D. students at Harvard University, T. E. Raghunathen and Leisa Weld, both of whom are completing theses on aspects of multiple imputation. This fourth and final version has benefitted from many suggestions from Raghunathen, Weld, Roderick J. A. Little and Alan Zaslavsky, and was facilitated by Raghunathen's computing help, and Bea Shube's and Rosalyn Farkas's editorial advice and patience. It too could have been postponed, waiting for improved results to come from ongoing research, but I believe the existing perspective is highly useful and that publication will stimulate new work. In fact, although many of the problems at the end of the chapters are rather standard exercises designed to check understanding of the material being presented, other problems involve issues that I consider research topics for term papers in a graduate-level course on survey methods or even points of departure for Ph.D. theses.

Since the summer of 1983, support for my work on multiple imputation and my graduate students' work at the University of Chicago and Harvard University has been primarily provided by a grant from NSF (SES-83-11428), and I am very grateful for this funding as well as additional support in 1986 from NSF (DMS-85-04332). The SES grant deals explicitly with the problem of the comparability of Census Bureau occupation and industry codes between 1970 and 1980, summarized here as Example 1.3. The creation of 1970 public-use files with multiply-imputed 1980 codes will be, I believe, an important milestone in the handling of missing values in public-use files.

This text is directed at applied survey statisticians with some theoretical background, but presents the necessary Bayesian and frequentist theory in the background Chapter 2. Chapter 3 derives, from the Bayesian perspective, general procedures for analyzing multiply-imputed data sets, and Chapter 4 evaluates the operating characteristics of these procedures from the randomization theory perspective. Particular procedures for creating multiple imputations are presented in Chapter 5 for cases with ignorable nonresponse and in Chapter 6 for cases with nonignorable nonresponse. Chapter 1 and the detailed table of contents are designed to allow the reader to obtain a rapid overview of the theory and practice of multiple imputation.

Multiple Imputation for Nonresponse in Surveys can serve as the basis for a course on survey methodology at the graduate level in a department of statistics, as I have done with earlier drafts at the University of Chicago and Harvard University. When utilized this way, I believe it should be supplemented with a more standard text, such as Cochran (1977), and readings from the National Academy of Sciences volumes on Incomplete Data (Madow et al., 1983).

I hope that the reader finds the material presented here to be a stimulating and useful contribution to the theory and practice of handling nonresponse in surveys.

DONALD B. RUBIN

Cambridge, Massachusetts
January 1987

Contents

TABLES AND FIGURES	xxiv
GLOSSARY	xxvii
1. INTRODUCTION	**1**
1.1. Overview	**1**
Nonresponse in Surveys	1
Multiple Imputation	2
Can Multiple Imputation Be Used in Nonsurvey Problems?	3
Background	4
1.2. Examples of Surveys with Nonresponse	**4**
Example 1.1. Educational Testing Service's Sample Survey of Schools	4
Example 1.2. Current Population Survey and Missing Incomes	5
Example 1.3. Census Public-Use Data Bases and Missing Occupation Codes	6
Example 1.4. Normative Aging Study of Drinking	7
1.3. Properly Handling Nonresponse	**7**
Handling Nonresponse in Example 1.1	7
Handling Nonresponse in Example 1.2	9
Handling Nonresponse in Example 1.3	10

	Handling Nonresponse in Example 1.4	10
	The Variety of Objectives When Handling Nonresponse	11
1.4.	**Single Imputation**	**11**
	Imputation Allows Standard Complete-Data Methods of Analysis to Be Used	11
	Imputation Can Incorporate Data Collector's Knowledge	12
	The Problem with One Imputation for Each Missing Value	12
	Example 1.5. Best-Prediction Imputation in a Simple Random Sample	13
	Example 1.6. Drawing Imputations from a Distribution (Example 1.5 continued)	14
1.5.	**Multiple Imputation**	**15**
	Advantages of Multiple Imputation	15
	The General Need to Display Sensitivity to Models of Nonresponse	16
	Disadvantages of Multiple Imputation	17
1.6.	**Numerical Example Using Multiple Imputation**	**19**
	Analyzing This Multiply-Imputed Data Set	19
	Creating This Multiply-Imputed Data Set	22
1.7.	**Guidance for the Reader**	**22**
	Problems	23

2. STATISTICAL BACKGROUND — 27

2.1.	**Introduction**	**27**
	Random Indexing of Units	27
2.2.	**Variables in the Finite Population**	**28**
	Covariates X	28
	Outcome Variables Y	29
	Indicator for Inclusion in the Survey I	29
	Indicator for Response in the Survey R	30
	Stable Response	30
	Surveys with Stages of Sampling	31

2.3. Probability Distributions and Related Calculations — 31
Conditional Probability Distributions — 32
Probability Specifications Are Symmetric in Unit Indices — 32
Bayes's Theorem — 33
Finding Means and Variances from Conditional Means and Variances — 33

2.4. Probability Specifications for Indicator Variables — 35
Sampling Mechanisms — 35
Examples of Unconfounded Probability Sampling Mechanisms — 37
Examples of Confounded and Nonprobability Sampling Mechanisms — 38
Response Mechanisms — 38

2.5. Probability Specifications for (X, Y) — 39
de Finetti's Theorem — 40
Some Intuition — 40
Example 2.1. A Simple Normal Model for Y_i — 40
Lemma 2.1. Distributions Relevant to Example 2.1 — 41
Example 2.2. A Generalization of Example 2.1 — 42
Example 2.3. An Application of Example 2.2: The Bayesian Bootstrap — 44
Example 2.4. Y_i Approximately Proportional to X_i — 46

2.6. Bayesian Inference for a Population Quantity — 48
Notation — 48
The Posterior Distribution for $Q(X, Y)$ — 48
Relating the Posterior Distribution of Q to the Posterior Distribution of Y_{nob} — 49
Ignorable Sampling Mechanisms — 50
Result 2.1. An Equivalent Definition for Ignorable Sampling Mechanisms — 50
Ignorable Response Mechanisms — 51
Result 2.2. Ignorability of the Response Mechanism When the Sampling Mechanism Is Ignorable — 51
Result 2.3. The Practical Importance of Ignorable Mechanisms — 52

	Relating Ignorable Sampling and Response Mechanisms to Standard Terminology in the Literature on Parametric Inference from Incomplete Data	53
2.7.	**Interval Estimation**	**54**
	General Interval Estimates	55
	Bayesian Posterior Coverage	55
	Example 2.5. Interval Estimation in the Context of Example 2.1	56
	Fixed-Response Randomization-Based Coverage	56
	Random-Response Randomization-Based Coverage	58
	Nominal versus Actual Coverage of Intervals	58
2.8.	**Bayesian Procedures for Constructing Interval Estimates, Including Significance Levels and Point Estimates**	**59**
	Highest Posterior Density Regions	59
	Significance Levels—p-Values	60
	Point Estimates	62
2.9.	**Evaluating the Performance of Procedures**	**62**
	A Protocol for Evaluating Procedures	63
	Result 2.4. The Average Coverages Are All Equal to the Probability That C Includes Q	64
	Further Comments on Calibration	64
2.10.	**Similarity of Bayesian and Randomization-Based Inferences in Many Practical Cases**	**65**
	Standard Asymptotic Results Concerning Bayesian Procedures	66
	Extensions of These Standard Results	66
	Practical Conclusions of Asymptotic Results	67
	Relevance to the Multiple-Imputation Approach to Nonresponse	67
	Problems	68
3.	**UNDERLYING BAYESIAN THEORY**	**75**
3.1.	**Introduction and Summary of Repeated-Imputation Inferences**	**75**
	Notation	75

	Combining the Repeated Complete-Data Estimates and Variances	76
	Scalar Q	77
	Significance Levels Based on the Combined Estimates and Variances	77
	Significance Levels Based on Repeated Complete-Data Significance Levels	78
	Example 3.1. Inference for Regression Coefficients	79
3.2.	**Key Results for Analysis When the Multiple Imputations Are Repeated Draws from the Posterior Distribution of the Missing Values**	**81**
	Result 3.1. Averaging the Completed-Data Posterior Distribution of Q over the Posterior Distribution of Y_{mis} to Obtain the Actual Posterior Distribution of Q	82
	Example 3.2. The Normal Model Continued	82
	The Posterior Cumulative Distribution Function of Q	83
	Result 3.2. Posterior Mean and Variance of Q	84
	Simulating the Posterior Mean and Variance of Q	85
	Missing and Observed Information with Infinite m	85
	Inference for Q from Repeated Completed-Data Means and Variances	86
	Example 3.3. Example 3.2 Continued	87
3.3.	**Inference for Scalar Estimands from a Modest Number of Repeated Completed-Data Means and Variances**	**87**
	The Plan of Attack	88
	The Sampling Distribution of $\mathbf{S}_m$ Given (X, Y_{obs}, R_{inc})	88
	The Conditional Distribution of $(\overline{Q}_\infty, \overline{U}_\infty)$ Given $\mathbf{S}_m$ and B_∞	89
	The Conditional Distribution of Q Given $\mathbf{S}_m$ and B_∞	89
	The Conditional Distribution of B_m Given $\mathbf{S}_m$	90
	The Conditional Distribution of $\overline{U}_m + (1 + m^{-1})B_\infty$ Given $\mathbf{S}_m$	90
	Approximation 3.1 Relevant to the Behrens-Fisher Distribution	91
	Applying Approximation 3.1 to Obtain (3.3.9)	92
	The Approximating t Reference Distribution for Scalar Q	92

	Example 3.4. Example 3.3 Continued	92
	Fraction of Information Missing Due to Nonresponse	93
3.4.	**Significance Levels for Multicomponent Estimands from a Modest Number of Repeated Completed-Data Means and Variance–Covariance Matrices**	**94**
	The Conditional Distribution of Q Given $\mathbf{S}_m$ and B_∞	94
	The Bayesian p-Value for a Null Value Q_0 Given $\mathbf{S}_m$: General Expression	95
	The Bayesian p-Value Given $\mathbf{S}_m$ with Scalar Q	95
	The Bayesian p-Value Given $\mathbf{S}_m$ with Scalar Q— Closed-Form Approximation	96
	p-Values with B_∞ *a Priori* Proportional to T_∞	96
	p-Values with B_∞ *a Priori* Proportional to T_∞— Closed-Form Approximation	97
	p-Values When B_∞ Is Not *a Priori* Proportional to $\bar{U}_\infty$	98
3.5.	**Significance Levels from Repeated Completed-Data Significance Levels**	**99**
	A New Test Statistic	99
	The Asymptotic Equivalence of $\tilde{D}_m$ and $\hat{D}_m$—Proof	100
	Integrating over r_m to Obtain a Significance Level from Repeated Completed-Data Significance Levels	100
3.6.	**Relating the Completed-Data and Complete-Data Posterior Distributions When the Sampling Mechanism Is Ignorable**	**102**
	Result 3.3. The Completed-Data and Complete-Data Posterior Distributions Are Equal When Sampling and Response Mechanisms Are Ignorable	103
	Using i.i.d. Modeling	104
	Result 3.4. The Equality of Completed-Data and Complete-Data Posterior Distributions When Using i.i.d. Models	104
	Example 3.5. A Situation in Which Conditional on θ_{XY}, the Completed-Data and Complete-Data Posterior Distributions of Q Are Equal—Condition (3.6.7)	105
	Example 3.6. Cases in Which Condition (3.6.7) Nearly Holds	105

	Example 3.7. Situations in Which the Completed-Data and Complete-Data Posterior Distributions of θ_{XY} Are Equal—Condition (3.6.8)	106
	Example 3.8. A Simple Case Illustrating the Large-Sample Equivalence of Completed-Data and Complete-Data Posterior Distributions of θ_{XY}	106
	The General Use of Complete-Data Statistics	106
Problems		107

4. RANDOMIZATION-BASED EVALUATIONS — 113

4.1. Introduction — 113

Major Conclusions — 113
Large-Sample Relative Efficiency of Point Estimates — 114
Large-Sample Coverage of t-Based Interval Estimates — 114
Outline of Chapter — 115

4.2. General Conditions for the Randomization-Validity of Infinite-m Repeated-Imputation Inferences — 116

Complications in Practice — 117
More General Conditions for Randomization-Validity — 117
Definition: Proper Multiple-Imputation Methods — 118
Result 4.1. If the Complete-Data Inference Is Randomization-Valid and the Multiple-Imputation Procedure Is Proper, Then the Infinite-m Repeated-Imputation Inference Is Randomization-Valid under the Posited Response Mechanism — 119

4.3. Examples of Proper and Improper Imputation Methods in a Simple Case with Ignorable Nonresponse — 120

Example 4.1. Simple Random Multiple Imputation — 120
Why Variability Is Underestimated Using the Multiple-Imputation Hot-Deck — 122
Example 4.2. Fully Normal Bayesian Repeated Imputation — 123
Example 4.3. A Nonnormal Bayesian Imputation Procedure That Is Proper for the Standard Inference—The Bayesian Bootstrap — 123
Example 4.4. An Approximately Bayesian yet Proper Imputation Method—The Approximate Bayesian Bootstrap — 124

	Example 4.5. The Mean and Variance Adjusted Hot-Deck	124
4.4.	**Further Discussion of Proper Imputation Methods**	**125**
	Conclusion 4.1. Approximate Repetitions from a Bayesian Model Tend to Be Proper	125
	The Heuristic Argument	126
	Messages of Conclusion 4.1	126
	The Importance of Drawing Repeated Imputations Appropriate for the Posited Response Mechanism	127
	The Role of the Complete-Data Statistics in Determining Whether a Repeated Imputation Method Is Proper	127
4.5.	**The Asymptotic Distribution of $(\bar{Q}_m, \bar{U}_m, B_m)$ for Proper Imputation Methods**	**128**
	Validity of the Asymptotic Sampling Distribution of S_m	128
	The Distribution of $(\bar{Q}_m, \bar{U}_m, B_m)$ Given (X, Y) for Scalar Q	129
	Random-Response Randomization-Based Justification for the t Reference Distribution	130
	Extension of Results to Multicomponent Q	131
	Asymptotic Efficiency of $\bar{Q}_m$ Relative to $\bar{Q}_\infty$	131
4.6.	**Evaluations of Finite-m Inferences with Scalar Estimands**	**132**
	Small-Sample Efficiencies of Asymptotically Proper Imputation Methods from Examples 4.2–4.5	132
	Large-Sample Coverages of Interval Estimates Using a t Reference Distribution and Proper Imputation Methods	134
	Small-Sample Monte Carlo Coverages of Asymptotically Proper Imputation Methods from Examples 4.2–4.5	135
	Evaluation of Significance Levels	135
4.7.	**Evaluation of Significance Levels from the Moment-Based Statistics D_m and $\tilde{D}_m$ with Multicomponent Estimands**	**137**
	The Level of a Significance Testing Procedure	138
	The Level of D_m—Analysis for Proper Imputation Methods and Large Samples	138
	The Level of D_m—Numerical Results	139

	The Level of $\tilde{D}_m$—Analysis	139
	The Effect of Unequal Fractions of Missing Information on $\tilde{D}_m$	141
	Some Numerical Results for $\tilde{D}_m$ with $k' = (k+1)\nu/2$	141
4.8.	**Evaluation of Significance Levels Based on Repeated Significance Levels**	**144**
	The Statistic $\hat{\tilde{D}}_m$	144
	The Asymptotic Sampling Distribution of $\bar{d}_m$ and s_d^2	144
	Some Numerical Results for $\hat{\tilde{D}}_m$	145
	The Superiority of Multiple Imputation Significance Levels	145
Problems		148
5.	**PROCEDURES WITH IGNORABLE NONRESPONSE**	**154**
5.1.	**Introduction**	**154**
	No Direct Evidence to Contradict Ignorable Nonresponse	155
	Adjust for All Observed Differences and Assume Unobserved Residual Differences Are Random	155
	Univariate Y_i and Many Respondents at Each Distinct Value of X_i That Occurs Among Nonrespondents	156
	The More Common Situation, Even with Univariate Y_i	156
	A Popular Implicit Model—The Census Bureau's Hot-Deck	157
	Metric-Matching Hot-Deck Methods	158
	Least-Squares Regression	159
	Outline of Chapter	159
5.2.	**Creating Imputed Values under an Explicit Model**	**160**
	The Modeling Task	160
	The Imputation Task	161
	Result 5.1. The Imputation Task with Ignorable Nonresponse	162
	The Estimation Task	163
	Result 5.2. The Estimation Task with Ignorable Nonresponse When $\theta_{Y|X}$ and θ_X Are *a Priori* Independent	164

Result 5.3. The Estimation Task with Ignorable Nonresponse, $\theta_{Y|X}$ and θ_X *a Priori* Independent, and Univariate Y_i 165

A Simplified Notation   165

5.3. Some Explicit Imputation Models with Univariate Y_i and Covariates   **166**

Example 5.1. Normal Linear Regression Model with Univariate Y_i 166

Example 5.2. Adding a Hot-Deck Component to the Normal Linear Regression Imputation Model   168

Extending the Normal Linear Regression Model   168

Example 5.3. A Logistic Regression Imputation Model for Dichotomous Y_i 169

5.4. Monotone Patterns of Missingness in Multivariate Y_i   **170**

Monotone Missingness in Y—Definition   171

The General Monotone Pattern—Description of General Techniques   171

Example 5.4. Bivariate Y_i and an Implicit Imputation Model   172

Example 5.5. Bivariate Y_i with an Explicit Normal Linear Regression Model   173

Monotone-Distinct Structure   174

Result 5.4. The Estimation Task with a Monotone-Distinct Structure   175

Result 5.5. The Imputation Task with a Monotone-Distinct Structure   177

5.5. Missing Social Security Benefits in the Current Population Survey   **178**

The CPS–IRS–SSA Exact Match File   178

The Reduced Data Base   179

The Modeling Task   179

The Estimation Task   180

The Imputation Task   181

Results Concerning Absolute Accuracies of Prediction   181

Inferences for the Average OASDI Benefits for the Nonrespondents in the Sample   184

Results on Inferences for Population Quantities   185

CONTENTS

5.6. Beyond Monotone Missingness — 186
Two Outcomes Never Jointly Observed—Statistical Matching of Files — 186
Example 5.6. Two Normal Outcomes Never Jointly Observed — 187
Problems Arising with Nonmonotone Patterns — 188
Discarding Data to Obtain a Monotone Pattern — 189
Assuming Conditional Independence Among Blocks of Variables to Create Independent Monotone Patterns — 190
Using Computationally Convenient Explicit Models — 191
Iteratively Using Methods for Monotone Patterns — 192
The Sampling/Importance Resampling Algorithm — 192
Some Details of SIR — 193
Example 5.7. An Illustrative Application of SIR — 194
Problems — 195

6. PROCEDURES WITH NONIGNORABLE NONRESPONSE — 202

6.1. Introduction — 202
Displaying Sensitivity to Models for Nonresponse — 202
The Need to Use Easily Communicated Models — 203
Transformations to Create Nonignorable Imputed Values from Ignorable Imputed Values — 203
Other Simple Methods for Creating Nonignorable Imputed Values Using Ignorable Imputation Models — 203
Essential Statistical Issues and Outline of Chapter — 204

6.2. Nonignorable Nonresponse with Univariate Y_i and No X_i — 205
The Modeling Task — 205
The Imputation Task — 206
The Estimation Task — 206
Two Basic Approaches to the Modeling Task — 207
Example 6.1. The Simple Normal Mixture Model — 207
Example 6.2. The Simple Normal Selection Model — 209

6.3. Formal Tasks with Nonignorable Nonresponse — 210
The Modeling Task—Notation — 210

Two General Approaches to the Modeling Task 211
Similarities with Ignorable Case 211
The Imputation Task 212
Result 6.1. The Imputation Task with Nonignorable Nonresponse 212
Result 6.2. The Imputation Task with Nonignorable Nonresponse When Each Unit Is Either Included in or Excluded from the Survey 212
The Estimation Task 213
Result 6.3. The Estimation Task with Nonignorable Nonresponse When $\theta_{YR|X}$ Is *a Priori* Independent of θ_X 213
Result 6.4. The Estimation Task with Nonignorable Nonresponse When $\theta_{Y|XR}$ Is *a Priori* Independent of $(\theta_{R|X}, \theta_X)$ and Each Unit Is Either Included in or Excluded from the Survey 213
Result 6.5. The Imputation and Estimation Tasks with Nonignorable Nonresponse and Univariate Y_i 214
Monotone Missingness 214
Result 6.6. The Estimation and Imputation Tasks with a Monotone-Distinct Structure and a Mixture Model for Nonignorable Nonresponse 214
Selection Modeling and Monotone Missingness 215

6.4. Illustrating Mixture Modeling Using Educational Testing Service Data **215**
The Data Base 216
The Modeling Task 216
Clarification of Prior Distribution Relating Nonrespondent and Respondent Parameters 217
Comments on Assumptions 218
The Estimation Task 219
The Imputation Task 219
Analysis of Multiply-Imputed Data 221

6.5. Illustrating Selection Modeling Using CPS Data **222**
The Data Base 223
The Modeling Task 224
The Estimation Task 225
The Imputation Task 225

	Accuracy of Results for Single Imputation Methods	226
	Estimates and Standard Errors for Average log(wage) for Nonrespondents in the Sample	227
	Inferences for Population Mean log(wage)	229
6.6.	**Extensions to Surveys with Follow-Ups**	**229**
	Ignorable Nonresponse	231
	Nonignorable Nonresponse with 100% Follow-Up Response	231
	Example 6.3. 100% Follow-Up Response in a Simple Random Sample of Y_i	232
	Ignorable Hard-Core Nonresponse Among Follow-Ups	233
	Nonignorable Hard-Core Nonresponse Among Follow-Ups	233
	Waves of Follow-Ups	234
6.7.	**Follow-Up Response in a Survey of Drinking Behavior Among Men of Retirement Age**	**234**
	The Data Base	235
	The Modeling Task	235
	The Estimation Task	235
	The Imputation Task	235
	Inference for the Effect of Retirement Status on Drinking Behavior	239
Problems		240

REFERENCES   244

AUTHOR INDEX   251

SUBJECT INDEX   253

Tables and Figures

Figure 1.1.	Data set with m imputations for each missing datum.	3
Table 1.1.	Artificial example of survey data and multiple imputation.	20
Table 1.2.	Analysis of multiply-imputed data set of Table 1.1.	21
Figure 2.1.	Matrix of variables in a finite population of N units.	29
Figure 2.2.	Contours of the posterior distribution of Q with the null value Q_0 indicated. The significance level of Q_0 is the posterior probability that Q is in the shaded area and beyond.	62
Table 4.1.	Large-sample relative efficiency (in %) when using a finite number of proper imputations, m, rather than an infinite number, as a function of the fraction of missing information, γ_0: RE $= (1 + \gamma_0/m)^{-1/2}$.	114
Table 4.2.	Large-sample coverage probability (in %) of interval estimates based on the t reference distribution, (3.1.8), as a function of the number of proper imputations, $m \geq 2$; the fraction of missing information, γ_0; and the nominal level, $1 - \alpha$. Also included for contrast are results based on single imputation, $m = 1$, using the complete-data normal reference distribution (3.1.1) with $\hat{Q}$ replaced by $\bar{Q}_1 = \hat{Q}_{*1}$ and U replaced by $\bar{U}_1 = U_{*1}$.	115
Table 4.3.	Simulated coverages (in %) of asymptotically proper multiple ($m = 2$) imputation procedures with nominal levels 90% and 95%, using t-based inferences, response rates n_1/n, and normal and nonnormal data (Laplace, lognormal $=\exp N(0, 1)$); maximum standard error $<1\%$.	136

Table 4.4.	Large-sample level (in %) of D_m with $F_{k,\nu}$ reference distribution as a function of nominal level, α; number of components being tested, k; number of proper imputations, m; and fraction of missing information, γ_0. Accuracy of results = 5000 simulations of (4.7.8) with ρ_0 set to I.	140
Table 4.5.	Large-sample level (in %) of $\tilde{D}_m$ with $F_{k,(k+1)\nu/2}$ reference distribution as a function of number of components being tested, k; number of proper imputations, m; fraction of missing information, γ_0; and variance of fractions of missing information, 0 (zero), S (small), L (large). Accuracy of results = 5000 simulations of (4.7.9).	142
Table 4.6.	Large-sample level (in %) of $\hat{\tilde{D}}_m$ with $F_{k,(1+k^{-1})\hat{\nu}/2}$ reference distribution as a function of number of components being tested, k; number of proper imputations, m; fraction of missing information, γ_0; and variance of fractions of missing information, 0 (zero), S (small), L (large). Accuracy of results = 5000 simulations of (4.7.7).	146
Table 4.7.	Large-sample level (in %) of d_{*1} with χ^2_k reference distribution as a function of nominal level α; number of components being tested, k; and fraction of missing information, γ_0.	147
Figure 5.1.	A monotone pattern of missingness, 1 = observed, 0 = missing.	171
Figure 5.2.	Artificial example illustrating hot-deck multiple imputation with a monotone pattern of missing data; parentheses enclose $m = 2$ imputations.	172
Table 5.1.	Multiple imputations of OASDI benefits for nonrespondents 62–71 years of age.	182
Table 5.2.	Multiple imputations of OASDI benefits for nonrespondents over 72 years of age.	183
Table 5.3.	Accuracies of imputation methods with respect to mean absolute deviation (MAD) and root mean squared deviation (RMS).	183
Table 5.4.	Comparison of estimates (standard errors) for mean OASDI benefits implied by imputation methods for nonrespondent groups in the sample.	184

Table 5.5.	Comparison of estimates (standard errors) for mean OASDI benefits implied by imputation methods for groups in the population.	185
Table 5.6.	Example from Marini, Olsen and Rubin (1980) illustrating how to obtain a monotone pattern of missing data by discarding data; 1 = observed, 0 = missing.	190
Table 6.1.	Summary of repeated-imputation intervals for variable 17B in educational example.	221
Table 6.2.	Background variables X for GRZ example on imputation of missing incomes.	223
Table 6.3.	Root-mean-squared error of imputations of log-wage: Impute posterior mean given θ fixed at MLE, $\hat{\theta}$.	226
Table 6.4.	Repeated-imputation estimates (standard errors) for average log(wage) for nonrespondents in the sample under five imputation procedures.	228
Figure 6.1.	Schematic data structure with follow-up surveys of nonrespondents: boldface produces Y data.	230
Table 6.5.	Mean alcohol consumption level and retirement status for respondents and nonrespondents within birth cohort: Data from 1982 Normative Aging Study drinking questionnaire.	236
Table 6.6.	Summary of least-squares estimates of the regression of log(1 + drinks/day) on retirement status (0 = working, 1 = retired), birth year, and retirement status × birth year interaction.	237
Table 6.7.	Five values of regression parameters for nonrespondents drawn from their posterior distribution.	237
Table 6.8.	Five imputed values of log(1 + drinks/day) for each of the 74 non-followed-up nonrespondents.	238
Table 6.9.	Sets of least-squares estimates from the five data sets completed by imputation.	239
Table 6.10.	Repeated-imputation estimates, standard errors, and percentages of missing information for the regression of log(1 + drinks/day) on retirement status, birth year, and retirement status × birth year interaction.	239

Glossary

Basic Random Variables

$X = N \times q$ matrix of fully observed covariates $= (X_{ij})$	28
$X_i = i$th row of $X =$ values of X for ith unit	28
$Y = N \times p$ matrix of partially observed outcome variables $= (Y_{ij})$	29
$Y_i = i$th row of $Y =$ values of Y for ith unit	29
$Y_{[j]} = j$th column of $Y = j$th outcome variable	171
$I = N \times p$ 0–1 indicator for inclusion of Y in survey $= (I_{ij})$	29
$I_i = i$th row of $I =$ indicator for outcomes included for unit i	29
$R = N \times p$ 0–1 indicator for response on $Y = (R_{ij})$	30
$R_i = i$th row of $R =$ indicator for response for ith unit	30

Index Sets Describing Portions of Y

$\text{inc} = \{(i, j) \mid I_{ij} = 1\} =$ included in survey	48
$\text{exc} = \{(i, j) \mid I_{ij} = 0\} =$ excluded from survey	48
$\text{obs} = \{(i, j) \mid I_{ij} R_{ij} = 1\} =$ observed	48
$\text{nob} = \{(i, j) \mid I_{ij} R_{ij} = 0\} =$ not observed	48
$\text{mis} = \{(i, j) \mid I_{ij}(1 - R_{ij}) = 1\} =$ missing (i.e., included but not observed)	48
$\text{inc}(i) = \{j \mid I_{ij} = 1\} =$ included in survey for unit i	162
$\text{exc}(i) = \{j \mid I_{ij} = 0\} =$ excluded from survey for unit i	162
$\text{obs}(i) = \{j \mid I_{ij} R_{ij} = 1\} =$ observed for unit i	162
$\text{nob}(i) = \{j \mid I_{ij} R_{ij} = 0\} =$ not observed for unit i	162
$\text{mis}(i) = \{j \mid I_{ij}(1 - R_{ij}) = 1\} =$ missing for unit i	162

$obs[j] = \{i | I_{ij} R_{ij} = 1\}$ = units with $Y_{[j]}$ observed   175
$nob[j] = \{i | I_{ij} R_{ij} = 0\}$ = units with $Y_{[j]}$ not observed   176
$mis[j] = \{i | I_{ij}(1 - R_{ij}) = 1\}$ = units with $Y_{[j]}$ missing   177
$ob = \cup_j obs[j]$ = units with observed values   165
$ms = \cup_j mis[j]$ = units with missing values   162

Complete-Data Statistics and Their Values in Multiply-Imputed Data Set

$\hat{Q}(X, Y_{inc}, I)$ = complete-data estimate of k-component estimand
$Q = Q(X, Y)$ 75

$U(X, Y_{inc}, I)$ = variance of $(Q - \hat{Q})$, $k \times k$ matrix   75

$\hat{Q}_{*l}$, $l = 1, \ldots, m$ = values of $\hat{Q}$ 76

U_{*l}, $l = 1, \ldots, m$ = values of U 76

d_{*l}, $l = 1, \ldots, m$ = values of χ^2 statistics testing $Q = Q_0$ 78

$\mathbf{S}_m = \{\hat{Q}_{*l}, U_{*l}, l = 1, \ldots, m\}$ 88

Repeated-Imputation Summary Statistics Used to Create Repeated-Imputation Inference for Estimand Q

$\overline{Q}_m = \sum_{l=1}^m \hat{Q}_{*l}/m$ = repeated-imputation estimate of Q, k-component row vector   76

$\overline{U}_m = \sum_{l=1}^m U_{*l}/m$, $k \times k$ matrix   76

$B_m = \sum_{l=1}^m (\hat{Q}_{*l} - \overline{Q}_m)'(\hat{Q}_{*l} - \overline{Q}_m)/(m-1)$, $k \times k$ matrix   76

$T_m = \overline{U}_m + (1 + m)^{-1} B_m$ = total variance of $(Q - \overline{Q}_m)$, $k \times k$ matrix   76

$r_m = (1 + m^{-1}) \text{Tr}(B_m \overline{U}_m^{-1})/k$ = fractional increase in variance due to nonresponse   78

$\nu = (m - 1)(1 + r_m^{-1})^2$ = degrees of freedom in reference distribution   77

$\gamma_m = \dfrac{r_m + 2/(\nu + 3)}{r_m + 1}$ = fraction of information missing due to nonresponse   77

$D_m = (Q_0 - \overline{Q}_m) T_m^{-1} (Q_0 - \overline{Q}_m)'/k$ = moment-based statistic testing $Q = Q_0$ 77

$\tilde{D}_m = (1 - r_m^{-1})(Q_0 - \overline{Q}_m) \overline{U}_m^{-1} (Q_0 - \overline{Q}_m)'/k$ = moment-based statistic testing $Q = Q_0$ 78

$\hat{D}_m$ = statistic testing $Q = Q_0$; asymptotically equivalent to $\tilde{D}_m$ 78
$\hat{\hat{D}}_m$ = test statistic approximating $\hat{D}_m$ using only $d_{*1}, \ldots, d_{*m}$ 79

Variances Relevant for Randomization-Based Evaluations, $k \times k$ matrices

$U_0 = U_0(X, Y) = E(U|X, Y) = V(\hat{Q}|X, Y)$ 118
$B_0 = B_0(X, Y) = E(B|X, Y)$ where $B = V(\overline{Q}_\infty|X, Y, I)$ 119
$T_0 = T_0(X, Y) = V(\overline{Q}_\infty|X, Y)$ 116
γ_0 = eigenvalues of B_0 with respect to T_0 = population fractions of missing information   131

CHAPTER 1

Introduction

1.1. OVERVIEW

In a census of a population, an attempt is made to collect information from each unit in the population. For example, in a census of the U.S. population, each person in the United States is contacted, and age, gender, years of education, and other characteristics are recorded.

In a sample survey of a population, the same sort of information is sought, but only some of the units in the population, those in the sample, are contacted. In well-designed sample surveys, the choice of the sample is carefully made in order to make inferences to the population both reliable and straightforward to obtain.

Nonresponse in Surveys

In many censuses and sample surveys, some of the units contacted do not respond to at least some items being asked. Such nonresponse, which we will call survey nonresponse whether it arises from a census or a sample survey, is common in practice whenever the population consists of units such as individual people, households, or businesses. The problem created by survey nonresponse is, of course, that data values intended by survey design to be observed are in fact missing. These missing values not only mean less efficient estimates because of the reduced size of the data base but also that standard complete-data methods cannot be immediately used to analyze the data. Moreover, possible biases exist because the respondents are often systematically different from the nonrespondents; of particular concern, these biases are difficult to eliminate since the precise reasons for nonresponse are usually not known.

An extended definition of survey nonresponse includes situations in which missing data arise from the processing of information provided by

1

units rather than from the refusal of units to provide information. For example, editing procedures may eliminate responses judged to be impossible (e.g., age equals 187 years), and restricted resources may limit the coding of open-ended responses to a subsample of units (e.g., although all units provide descriptions of their occupations, only a subsample of their responses are read and coded by survey staff). An even more extended definition of survey nonresponse includes any situation in which there are missing values in the rectangular units-by-variables data matrix to be analyzed, even if no attempt was made to record some of the missing values (e.g., income questions are only asked of a subsample of the units in the survey, or ages of children are only recorded to the nearest year but for analysis are needed to the nearest month). Multiple imputation is relevant to all such problems of deficient data, and so the broadest definition of survey nonresponse is accepted here.

Multiple Imputation

Multiple imputation is the technique that replaces each missing or deficient value with two or more acceptable values representing a distribution of possibilities; this idea was originally proposed by Rubin (1977b, 1978a). Figure 1.1 depicts a multiply-imputed data set, where each missing datum is replaced by a pointer to a vector of m values. The m values are ordered in the sense that the first components of the vectors when substituted for the missing values result in one data set, the second components result in a second data set, and so on. The imputed values are stored in an auxiliary matrix with one row for each missing value and m columns. In common practice it would be inconvenient to have an auxiliary matrix of imputations much larger than the original data matrix, and consequently, a multiply-imputed data set is most useful to practice when the fraction of values missing is not excessive and when m is modest, say between 2 and 10.

In this text we will:

1. Describe how to generate multiple imputations in general and offer specific examples of multiple-imputation procedures that are easy to use, easy to communicate, and provide valid inferences.
2. Show how to draw inferences from a multiply-imputed data set.
3. Justify the resultant inferences by showing the senses in which they are valid.
4. Present several exploratory applications of multiple imputation to real data.

OVERVIEW

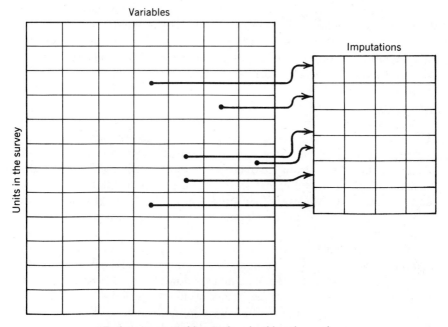

Each row vector of imputations is of length m, where
Model for first imputation = $\cdots$
Model for second imputation = $\cdots$
$$\vdots$$
Model for mth imputation = $\cdots$

Figure 1.1. Data set with m imputations for each missing datum.

Can Multiple Imputation Be Used in Nonsurvey Problems?

Multiple imputation can be used to handle missing data in nonsurvey contexts, but there are several reasons why the technique seems to be especially valuable in the survey context. First, surveys quite often collect very large amounts of data that will be analyzed by many users; consequently, there is often a desire to "fix-up" the data base (e.g., fill in the missing values) before sending it off for general use. Second, the analyses of many surveys rely on finely tuned, highly specific estimators and adjustments, often based on past experience with similar surveys; the existence of missing values means that the standard procedures cannot be automatically used, and corresponding procedures that appropriately adjust for the missing data may not be easy to derive. Third, it is rare that the missing values occur completely at random, as they might in some experimental contexts; in surveys, it is often reasonable to suspect that nonrespondents systematically differ from respondents, and thus it is desirable not only to adjust for nonresponse but also to study the effects of various assumed differences between respondents and nonrespondents. Fourth, imputation is common in survey practice, and multiple imputations can often be generated from simple modifications of existing single-imputation methods such as the Census Bureau hot-deck or regression methods.

Background

The previous literature on the theory and practice of handling missing survey data will not be reviewed here. The National Academy of Sciences Panel on Incomplete Data has recently produced three volumes, including an annotated bibliography, that serve this purpose well (Madow, Nisselson, and Olkin, 1983; Madow, Olkin, and Rubin, 1983; Madow and Olkin, 1983.) Here, we focus on developing the theory and practice of multiple imputation.

We assume that the reader is familiar with basic theory and practice of sample surveys based on randomization inference, such as presented in Cochran (1977), Hansen, Hurwitz, and Madow (1953), and Kish (1965). In particular, we assume that the reader would know how to analyze the survey under consideration if there were no nonresponse. We also assume that the reader is comfortable with basic Bayesian calculations and terminology, such as at the level of the introductory chapters of Box and Tiao (1973). The use of Bayesian inference in sample surveys is less well known to survey practitioners than randomization inference, although evidently used before randomization inference in real surveys. For example, in 1780 Laplace used Bayesian methods to estimate the population of France (Cochran, 1978).

1.2. EXAMPLES OF SURVEYS WITH NONRESPONSE

In order to understand the problem that multiple imputation is supposed to solve, and to judge how well it does so, it is important to obtain a feeling for a range of examples of survey nonresponse, as well as to see what it ideally means to "handle nonresponse" in these examples and the shortcomings of standard methods typically in use. Multiple imputation has been applied, at least in exploratory exercises, to all examples presented here.

Example 1.1. Educational Testing Service's Sample Survey of Schools

In 1971 Educational Testing Service (ETS) conducted a sample survey of 660 schools for the purpose of studying their compensatory reading programs. Although some background information on all schools was available from Census Bureau records (e.g., median income in ZIP code area), the critical information on types of compensatory reading programs and achievement levels of students in the schools was to be obtained from a questionnaire sent to the principals. By the end of the sample survey, only 472 of the 660 principals had returned this questionnaire, even though all had been mailed several reminders and had been contacted by telephone.

Nonresponse on this questionnaire is called unit nonresponse, in contrast to item nonresponse, because the unit (i.e., the nonresponding principal) refused to respond to any items on the survey instrument. If the principals had responded to some questionnaire items but left others unanswered, then we would say that the survey had suffered from item nonresponse rather than unit nonresponse.

Since the principals knew that the purpose of the survey was to study their compensatory reading programs, concern developed that the 188 nonresponding principals were systematically different from the 472 responding principals, perhaps having students with more severe reading problems or, perhaps, having reading programs that were less effective. In fact, on the basis of the census background information, it did appear that the respondents differed systematically from nonrespondents on observed relevant characteristics. Perhaps they also differed systematically on unobserved relevant characteristics. If the purpose of the survey had been only to select a few exemplary programs for study, this potential bias might have been of little concern, but one purpose of the survey was to describe the population of programs and their typical cost effectiveness. Consequently, the high nonresponse rate (188 out of 660) was considered to be a serious problem.

ETS had to try to ensure that its analyses of the data did not misrepresent the characteristics of the population of compensatory reading programs. For example, ETS wanted to produce estimates for "the average number of hours per week of compensatory reading that is offered" that reflected the increased uncertainty arising from the high nonresponse rate.

Example 1.2. Current Population Survey and Missing Incomes

The Current Population Survey (CPS), conducted by the Census Bureau, includes approximately 50,000 households monthly and is used to gather a variety of information. One type of information that has traditionally been relatively quite difficult to obtain and is becoming even more difficult to obtain is income data: Individuals are not particularly anxious to divulge their incomes, whether total, or amounts in certain categories such as social security benefits. The current 15–20% nonresponse rate on many income items does not just result in reduced efficiency of estimation for users of the

CPS. The more serious problem concerns bias. It is certainly possible that those refusing to answer income questions systematically differ from those willing to supply it. In fact, there exists evidence to suggest that middle-income people are more likely to be income respondents than low- or high-income people. Low-income individuals apparently tend to be "general nonrespondents" in that they are relatively likely to refuse to respond to many items (i.e., tend to generate unit nonresponse) whereas high-income individuals tend to be "specific nonrespondents" in that they are relatively likely to respond to most items, except income items for which they have a very high nonresponse rate (e.g., over 30%). Nonresponse to income items in the CPS, especially among high-income individuals, is an example of item nonresponse. That is, each unit included in the survey produces information on most items, but some units fail to provide information on some particular items. Subjects that refuse to participate in the CPS exhibit unit nonresponse.

Since the CPS is a major source of data for economists and other social scientists, as well as for businesses, it is important that realistic answers can be obtained from the CPS data bases in spite of the problems arising from income nonresponse.

Example 1.3. Census Public-Use Data Bases and Missing Occupation Codes

In 1980, the Census Bureau substantially modified its coding of the descriptions of the occupations held by individuals. An important consequence of this change is that public-use data bases from the 1980 Census have occupation codings that are not directly comparable to those on public-use data bases from previous censuses, in particular, 1970 public-use data bases. Estimated costs for double-coding 1970 public-use data bases (i.e., for coding all units on 1970 public-use data bases according to the new 1980 occupation coding system in addition to their existing occupation codes) are in the millions of dollars. There exists, however, a double-coded sample of 120,000 units from the 1970 Census, that is, with both 1980 and 1970 codes for all 120,000 units. Consequently, we can think of the lack of both codes for all units on 1970 public-use data bases as an enormous nonresponse problem: for 120,000 units in the 1970 Census public-use data bases, both old and new occupation codes are observed, whereas for all remaining units, only the old code is observed. Viewed this way, nonresponse on 1980 occupation codes in the 1970 public-use data bases is an example of item nonresponse.

The lack of a common coding system is considered by many economists and sociologists to be a very serious problem. If pre-1980 public-use data bases have one coding and the 1980 data bases have another coding, it will be very difficult to study such topics as occupational mobility and labor force shifts by demographic characteristics. Specific questions that would be difficult to address without 1980 codes on 1970 public-use data bases concern 1970–1980 shifts in occupational status of jobs held by males and females or by whites and nonwhites. Such questions are important because they address the issue of equal opportunity employment. Consequently, there is a need to supply both codes on 1970 public-use data bases.

Example 1.4. Normative Aging Study of Drinking

In 1982, a survey was conducted by the Veterans Administration in Boston to investigate drinking behavior in men aged 50–70. A total of 1423 men were mailed background and drinking behavior questionnaires: 1272 provided essentially complete information; 112 provided background information but were nonrespondents on the drinking behavior questionnaire; and 39 never even acknowledged receipt of the questionnaires. In 1983, 38 of the 112 nonrespondents on drinking behavior were interviewed, and complete information on the drinking behavior questionnaire was obtained. The primary objective of the study was to relate drinking behavior to retirement status and age, as by a linear regression model.

This example is similar to Example 1.1 in that potential nonresponse bias is an issue and the primary concern is with a realistic analysis rather than the production of public-use data bases. A new feature, however, is the availability of some followed-up nonrespondents. Although their drinking behavior data was collected under different circumstances than the respondents' data, there exists the possibility of using the follow-ups' data to adjust, at least partially, for potential systematic differences between respondents and nonrespondents.

1.3. PROPERLY HANDLING NONRESPONSE

Handling nonresponse means somewhat different things in these examples because the purposes of the data collection efforts and the kinds of nonresponse are not the same. Current practice is not generally adequate.

Handling Nonresponse in Example 1.1

In the ETS survey of schools, the data will be analyzed only by ETS for its final report. An appropriate method for dealing with the nonresponse satisfies three objectives: First, it adjusts estimates for the fact that on measured background variables the nonrespondents differ from respondents; second, it expands standard errors of estimates to reflect both the reduced sample size from 660 to 472 and the observed differences between nonrespondents and respondents; and third, it exposes sensitivity of estimates and standard errors to possible differences between nonrespondents and respondents on unmeasured background variables. The first two objectives are often not satisfactorily addressed in current practice with such data. The third objective is usually entirely ignored even though it is critical with data sets in which the reasons for nonresponse are not precisely understood.

There exists a variety of methods that practitioners tend to use with such data. One common practice would simply discard the 188 schools with incomplete data; this obviously will bias results if reasons for nonresponse are correlated with values of variables. For example, if the nonrespondents have lower-achieving students, analyses based on the respondents alone will overestimate the typical achievement of students in compensatory reading programs. A standard technique for improving the estimates based on responding units is to weight the responding units' data to compensate for the nonresponding units' missing data. Thus, in the ETS example, a responding school that is similar to two nonresponding schools with respect to the measured background variables would receive weight $1 + 2 = 3$, and so forth, where the total weight for the 472 responding schools would equal 660, the original sample size. This method implicitly assumes no nonresponse bias beyond that explained by the measured background variables. Furthermore, the method's apparent simplicity disappears with multivariate outcomes and item nonresponse since each unit then has, in principle, a different weight for each item. Thus, weighting adjustments tend to be confined to problems of unit nonresponse.

Many users of survey data confronted with item or unit nonresponse simply fill in the missing data. For example, the mean for each variable based on the 472 responding schools might be filled in for the missing values of the corresponding variable. Such a procedure, though easy, can easily bias results. First, the means of the filled-in variables are the same as the respondents' means, which may not be appropriate for nonrespondents because of nonresponse bias. Second, relationships among variables—for instance, as measured by correlations—can be badly distorted even in the absence of nonresponse bias.

A final *ad hoc* procedure often used when analyses are based on estimated means, variances, and correlations is to estimate the mean and variance of a variable using all units responding to that variable, and to estimate the correlation between two variables using all units responding to both variables. Although such a method might seem sensible, as with the other *ad hoc* procedures, it too can lead to biased results. For example, the resultant mean and variance are still truly appropriate only for respondents; also, a negative-definite correlation matrix can result, which would lead to negative eigenvalues in a principal components analysis, or negative variances in a regression analysis.

Handling Nonresponse in Example 1.2

The second example, nonresponse on income items in the CPS, exhibits a serious problem not present in the first example. The Census Bureau creates public-use data bases from the CPS, and these are released to many users with varying degrees of statistical sophistication. The Bureau, being the producer of these data bases, has an obligation to try to ensure that typical users can analyze these data bases with the standard statistical methods at their disposal, and moreover, that the resultant answers will not be misleading because of editing methods employed by the Bureau.

Current practice for handling missing income data in the CPS is to use a rather complicated "hot-deck" procedure for imputation, that is, for estimating and filling in the missing values. A hot-deck procedure finds for each nonrespondent a matching respondent, where matching means close with respect to variables observed for both. The CPS hot-deck employs many variables and modifies the coarseness of the matching variables depending on the ability to find matches. For example, "region of country" has five levels, but if no match can be found, it might be modified to have only two levels (e.g., North versus South). Even though this hot-deck imputation allows the Bureau to produce public-use data bases that can be analyzed by complete-data methods, the results of such analyses systematically underestimate variability because they treat imputed values as if they were known with certainty; this underestimation of variability might be serious in some cases since the nonresponse rate can be as high as 30% for some types of subjects. It is not obvious how to adjust the analyses to correct this problem. Furthermore, the reasons for nonresponse are not fully understood, and yet there is no indication of the sensitivity of inference to different assumptions about differences between nonrespondents and respondents. Nevertheless, the hot-deck is a standard Census Bureau tool, even being used, for example, in the Decennial Census to estimate the number of individuals living in unclassified households for the purpose of allocating congressional seats to the states. It would be important for practice if the hot-deck could be modified to properly handle nonresponse in the CPS.

Handling Nonresponse in Example 1.3

In Example 1.3, the reason for nonresponse on 1980 occupation codes on 1970 public-use data bases is precisely known—the Census Bureau decided in 1970 to record 1970 codes rather than 1980 codes. Consequently, in this example, there is no need to be concerned with sensitivity to different models of the reasons for nonresponse, which was an objective with Examples 1.1 and 1.2. There is, however, a serious problem that arises from the need to create public-use data bases: the public-use data bases are more than 10 times the size of the double-coded sample, which implies a nonresponse rate of over 90% on 1980 codes!

Two distinct options were initially proposed to handle missing 1980 occupation codes. The first option is to eliminate the missing data by spending the money to double-code the 1970 public-use data bases. The second option is to use some form of imputation. A hot-deck procedure and logistic regression modeling method, both based on the double-coded sample of 120,000 units, have been considered. The hot-deck imputation would find, for each 1970 unit without a 1980 occupation code, a unit in the double-coded sample that has similar values of variables recorded for both units, such as 1970 occupation code, income, gender, and age. The matching unit's observed 1980 occupation code is then imputed (i.e., filled in) for the missing 1980 value. Logistic regression imputation would build a model from the double-coded sample that would predict 1980 occupation code from 1970 occupation code and other predictors such as income, gender, and age, and then apply this model to the units with predictors observed but 1980 occupation code missing, in order to impute the most likely occupation. Although producing a filled-in data base by such imputation methods allows the typical user to employ standard complete-data methods of analysis, the results of such analyses systematically underestimate variability because they treat imputed values as if they were real.

Handling Nonresponse in Example 1.4

The issues created by the nonresponse in the Normative Aging Study are similar to those created in the ETS survey of schools (Example 1.1), with the additional feature that the follow-ups provide an opportunity for a data-based adjustment of differences between respondents and nonrespondents. If the follow-ups are regarded as a random sample of the nonrespondents, simple estimands such as the population mean can be handled using special double-sampling estimation techniques such as described in Cochran (1977, Chapter 12). More complicated estimands, such as regression coefficients, or analyses when it is not assumed that the follow-ups are a random sample from the nonrespondents, however, are generally not appropriately handled by standard procedures.

The Variety of Objectives When Handling Nonresponse

In order to handle the variety of types of nonresponse problems just described, we require a technique that has at least the following capabilities. First, it should allow standard complete-data methods to be used. Second, it should be capable of yielding valid inferences that (a) produce estimates that adjust for observed differences between respondents and nonrespondents and (b) produce standard errors that reflect the reduced sample size as well as the adjustment for the observed respondent–nonrespondent differences. Third, it should display the sensitivity of inferences to various plausible models for nonresponse. Current practice for handling nonresponse does not satisfy all of these objectives.

1.4. SINGLE IMPUTATION

Single imputation, that is, filling in a value for each missing value, is probably the most common method for handling item nonresponse in current survey practice. There are two major attractive features supporting this practice. First, standard complete-data methods of analysis can be used on the filled-in data set. Second, in the context of public-use data bases, the possibly substantial effort required to create sensible imputations need be carried out only once, by the data producer, and these imputations can incorporate the data collector's knowledge. Just as these two advantages are rather obvious and important, there are equally obvious and important disadvantages of single imputation: the single value being imputed can reflect neither sampling variability about the actual value when one model for nonresponse is being considered nor additional uncertainty when more than one model is being entertained. Both the advantages and disadvantages of single imputation deserve further comment.

Imputation Allows Standard Complete-Data Methods of Analysis to Be Used

The first major advantage of imputation is that once the values have been filled in, standard complete-data methods of analysis can be used. In contrast, some mathematical statistical approaches to nonresponse require new and possibly specialized computer programs in order to handle the nonresponse problem.

From a practitioner's point of view, it may not be at all obvious that the extra expense and effort required to design, write, and validate special computer programs is really worthwhile, say relative to field efforts aimed at reducing nonresponse or increasing sample size. The practical question is whether the answers under new theoretical models are really that much better than (or even as good as) the answers that can be found following straightforward imputation procedures.

It is easy to be sympathetic toward the imputation position because much practice and experience has verified many standard data analysis techniques as being appropriate in the absence of nonresponse, especially in particular contexts. Minimally, general acceptance of statistical conclusions and general understanding of substantive conclusions may be lost when these standard methods of data analysis are not used. Also, statistically unsophisticated users are probably more likely to reach reasonable conclusions when using familiar statistical techniques rather than sophisticated models beyond their understanding. Consequently, there is substantial practical advantage if imputation methods can be made theoretically sound.

Imputation Can Incorporate Data Collector's Knowledge

The second major advantage of imputation is that in many cases the imputations can be created just once by the data collector who may have much better information about and understanding of the process that creates nonresponse than the typical user. Such concerns are especially important in cases (such as Examples 1.2 and 1.3) in which the data collector (i.e., the Census Bureau) has more information at its disposal for imputation (some of it protected by confidentiality constraints) than will be available on the resultant public-use data bases, as well as greater resources for analysis than the typical consumer of the resultant data. Consequently, it is possible that data analysts, even those with a full arsenal of modern statistical tools, might reach better inferences by trusting the data collector's imputations than by applying sophisticated statistical models to a less rich data base. Provided imputations are flagged, the data collector's imputations can always be ignored if an explicit model for nonresponse is being considered by the statistically sophisticated user. Certainly, statistically unsophisticated users would quite often fare better using the data collector's imputations rather than using their own quick-fix imputations such as "fill in the mean."

The Problem with One Imputation for Each Missing Value

The obvious problem with imputation is that the missing values are not known, and yet the automatic application of complete-data methods to

SINGLE IMPUTATION

imputed data sets treats missing values as if they were known. Because of this, even if the mechanism creating nonresponse is perfectly understood as in Example 1.3, inferences based on the imputed data set will be too sharp since the extra variability due to the unknown missing values is not being taken into account. Also, quantities such as correlations that depend on variabilities can be badly biased. Furthermore, when nonresponse is not really understood, no account is being taken of the uncertainty arising from not knowing which nonresponse models for imputation are appropriate.

Example 1.5. Best-Prediction Imputation in a Simple Random Sample

In a simple random sample with no covariates, standard complete-data inferences follow from the statement that

$$(\bar{y} - \bar{Y}) \text{ is approximately normally distributed} \\ \text{with mean zero and variance } s^2(n^{-1} - N^{-1}) \quad (1.4.1)$$

where
n = sample size
$\bar{y}$ = sample mean
s^2 = sample variance
N = population size
$\bar{Y}$ = population mean.

See, for example, Cochran (1977, Chapter 2).

Suppose because of random nonresponse, only n_1 of the n values of Y are actually observed where $\bar{y}_1$ and s_1^2 are the sample mean and variance of the n_1 observed values. Because $\bar{y}_1$ and s_1^2 are essentially the sample mean and sample variance in a simple random sample of size n_1, under the usual normal assumptions, standard inferences for $\bar{Y}$ can be based on the statement that

$$(\bar{y}_1 - \bar{Y}) \text{ is approximately normally distributed} \\ \text{with mean zero and variance } s_1^2(n_1^{-1} - N^{-1}). \quad (1.4.2)$$

Suppose that instead of using (1.4.2), the best-prediction values of the missing Y are imputed, and (1.4.1) is applied with no distinction being made for the differing status of observed and imputed values. Under assumptions being made, the best prediction of each missing Y is the observed sample mean $\bar{y}_1$. Hence, the mean of all n values is $\bar{y}_1$, and the sample variance is $s_1^2(n_1 - 1)/(n - 1)$, since all missing Y are imputed equal to the sample mean. As a result, the variance of $(\bar{y} - \bar{Y})$ in (1.4.1) based on the data set completed by best-prediction imputation is $s_1^2(n^{-1} - N^{-1})(n_1 - 1)/(n - 1)$, which in comparison to the actual variance of $\bar{y}_1$ given in (1.4.2), is too small, essentially by a factor of $(n_1/n)^2$ for large n_1 and N/n_1. Thus, resulting interval estimates of $\bar{Y}$ will be too short by a factor approximately equal to n_1/n, leading to potentially severe undercoverage, and test statistics will be too large, leading to excessively large significance levels associated with null values of $\bar{Y}$. There are two reasons for these problems. First, we are pretending that the sample size is n when it really is n_1, and second, we are underestimating the variance by a factor, approximately equal to n_1/n, because we are imputing all $n - n_1$ values at the sample mean.

In general, it is intuitively clear that imputing the best prediction for each missing value must underestimate variability. Furthermore, in general, imputing the best value will not necessarily lead to correctly centered inferences unless the complete-data statistic is a linear function of the missing values. For example, consider estimating the population variance of Y; when best predictions have been imputed for each missing Y, the estimate based on the n values of Y is too small by the factor $(n_1 - 1)/(n - 1)$.

Example 1.6. Drawing Imputations from a Distribution
(Example 1.5 continued)

Suppose that in an attempt to avoid the underestimation of variance created by best-prediction imputation, the imputed values are drawn so that on average the mean of the imputed values is $\bar{y}_1$, and on average the sample variance of all n values is s_1^2. This result is nearly accomplished by drawing the missing Y randomly from the observed values. Treating the observed data and design as fixed, $\bar{y}$ and s^2 in (1.4.1) are random variables, where it is easy to show that when the imputed values are drawn with replacement, the expected values of $\bar{y}$ and s^2 are $\bar{y}_1$ and $s_1^2(1 - n_1^{-1}) \times [1 + n_1 n^{-1}(n - 1)^{-1}] \doteq s_1^2$, respectively (see Problems 11 and 12).

Even though the imputed values approximately preserve the observed distribution of the data, the variance of the reference distribution for $(\bar{y} - \bar{Y})$ in (1.4.1), that is, $s^2(n^{-1} - N^{-1})$, is incorrect. First, suppose the imputed values were drawn so that $\bar{y}$ always equaled $\bar{y}_1$ and s^2 always equaled s_1^2; the associated variance of $(\bar{y} - \bar{Y}) = (\bar{y}_1 - \bar{Y})$ should be $s_1^2(n_1^{-1} - N^{-1})$ as in (1.4.2) rather than $s^2(n^{-1} - N^{-1}) = s_1^2(n^{-1} - N^{-1})$, which is too small by at least the factor n_1/n, and so this procedure also leads to interval estimates that are too short and test statistics that are too big. Second, if the values are imputed with some randomization, (i) the actual variance of $\bar{y}$ is larger than the variance of $\bar{y}_1$, and (ii) s^2 varies about s_1^2, with the result that even $s_1^2(n_1^{-1} - N^{-1})$ underestimates the actual variance of $\bar{y}$ based on the randomly imputed data. A related problem with random drawing is that the efficiency of $\bar{y}$ is less than that of y_1; for example, when drawing imputed values with replacement, we have that the variance of $\bar{y}$ is larger than the variance of $\bar{y}_1$ by the factor $[1 + (1 - n_1/n)(1 - n_1/N)^{-1}(n_1 - 1)/n]$.

Although values of missing Y could be systematically chosen to give $\bar{y} = \bar{y}_1$ and $s^2 = s_1^2 n/n_1$, at least if $n - n_1 > 1$, such a pathway is hopeless for general practice: Some distributional characteristics are being distorted, and so some population parameters (e.g., the population variance, or with two variables, the population correlation) will not be properly estimated by standard complete-data analyses.

1.5. MULTIPLE IMPUTATION

Multiple imputation retains the virtues of single imputation and corrects its major flaws. The idea behind multiple imputation is that for each missing datum we impute several values, say m, instead of just one, as already displayed in Figure 1.1. These m values are ordered in the sense that the first set of values imputed for the missing values are used to form the first completed data set, and so on. Thus, the m imputations for each missing datum create m complete data sets. Each completed data set is analyzed using standard complete-data procedures just as if the imputed data were the real data obtained from the nonrespondents; commonly, such an analysis ignores the distinction between respondents and nonrespondents. Multiple imputation using modest m, say between 2 and 10, is designed for situations with a modest fraction of information missing due to nonresponse. Missing information is defined precisely in Chapter 3, but basically it measures the increase in variance of estimation due to nonresponse, and is determined by response rates and the ability of observed values to predict missing values successfully. Examples 1.1–1.4 all exhibit relatively modest fractions of missing information for most estimands of interest under reasonable models for nonresponse, even Example 1.3 because the missing 1980 codes are relatively well predicted by the observed values of 1970 codes, income, gender, age, etc.

Advantages of Multiple Imputation

Multiple imputation shares with single imputation the two basic advantages already mentioned, namely, the ability to use complete-data methods of analysis and the ability to incorporate the data collector's knowledge. In fact, the second basic advantage is not only retained but enchanced because multiple imputations allow data collectors to use their knowledge to reflect uncertainty about which values to impute. This uncertainty is of two types: sampling variability assuming the reasons for nonresponse are known and variability due to uncertainty about the reasons for nonresponse. Under each posited model for nonresponse, two or more imputations are created to reflect sampling variability under that model; imputations under more than one model for nonresponse reflect uncertainty about the reasons for nonresponse.

There exist three extremely important advantages to multiple imputation over single imputation. First, when imputations are randomly drawn in an attempt to represent the distribution of the data, as illustrated in Example 1.6, multiple imputation increases the efficiency of estimation. Specific results on the increase in efficiency are given in Chapter 4.

The second distinct advantage of multiple imputation over single imputation is that when the multiple imputations represent repeated random draws under a model for nonresponse, valid inferences—that is, ones that reflect the additional variability due to the missing values under that model—are obtained simply by combining complete-data inferences in a straightforward manner. The procedure is illustrated in Section 1.6 and presented in general in Chapter 3. Because of the simplicity of this technology, statistically unsophisticated users can reach valid inferences using only familiar complete-data tools. In addition, even when explicit models not considered by the data collector are to be applied by the statistically sophisticated data user, it may be computationally more efficient to draw inferences from these explicit models by using multiple imputation to simulate the correct inference than to carry out specialized mathematical analyses or to write specialized computer programs perhaps involving numerical integration.

The third distinct advantage of multiple imputation is that by generating repeated randomly drawn imputations under more than one model, it allows the straightforward study of the sensitivity of inferences to various models for nonresponse simply using complete-data methods repeatedly. This is an endeavor that the statistically unsophisticated user especially is extremely unlikely to consider in the absence of multiple imputation; yet it is very important because, in general, the analyses are reliant on assumptions unassailable by the observed data.

The General Need to Display Sensitivity to Models of Nonresponse

For an artificial example exhibiting the need to rely on external assumptions about the process that creates nonresponse, that is, assumptions external to the survey data, suppose that one-half of the sampled units are nonrespondents, and the respondents' values appear to be exactly the right half of a normal distribution. Which is a more appropriate estimate of the population mean: the sample mean for respondents or the minimum observed value for respondents? If the mean values for respondents and nonrespondents in the population were known to be the same, the sample mean would be more appropriate, whereas if nonresponse occurs only for smaller values of the variable *and* the values in the population are known to be approximately symmetrically distributed, the minimum observed value would be more appropriate.

The observed values for respondents cannot discriminate between these alternatives, and consequently inferences for the population mean will necessarily be sensitive to the choice of models for the process that creates nonresponse. Note that even if we assume nonresponse occurs only for the bottom 50% of the population values, the minimum observed value may not be a good estimate of the population mean: in large samples, the minimum observed value will nearly equal the population median, which does not equal the population mean without further assumptions about the distribution of values in the population, such as symmetry. In general, it is impossible to estimate a population quantity such as the mean without making assumptions about either the distribution of values for nonrespondents or the process that creates nonresponse. Such assumptions should ideally be based on familiarity with the content or the subject area of the survey and psychological aspects of the units of the survey and the survey instrument; recent books relevant to this subject include Bradburn and Sudman (1979), Dijkstra and Vander Zouwen (1982), Dillman (1978), Groves and Kahn (1979), Rosenthal and Rosnow (1975), and Sudman and Bradburn (1974).

If, in a particular survey without follow-up response, there is no single accepted class of assumptions about nonresponse, then it is obviously prudent to perform data analyses under a variety of plausible models for nonresponse. If (a) inferences vary in important ways as the models change and (b) the data cannot eliminate some models as inappropriate, then the tautological conclusion must be that the data cannot support sharp inferences without further specification of the models. A recent examination of the sensitivity of inferences to models for nonresponse in a particular survey is presented in Heitjan and Rubin (1986).

Disadvantages of Multiple Imputation

There are three obvious disadvantages of multiple imputation relative to single imputation. First, more work is needed to produce multiple imputations than single imputations. Second, more space is needed to store a multiply-imputed data set. Third, more work is needed to analyze a multiply-imputed data set than a singly-imputed data set. These disadvantages are not serious when m is modest. Modest m is adequate when fractions of missing information are modest. When fractions of missing information are large, modest-m multiple imputation is not fully satisfactory, but then single imputation can be disastrous.

With respect to the first disadvantage, often the multiple-imputation version of an existing single-imputation scheme is not difficult to implement. The artificial example in Section 1.6 and the examples in later chapters illustrate this point. Also, these examples show that in many cases it is not difficult to implement new multiple-imputation procedures using standard statistical programs.

With respect to the second disadvantage, in order to store a multiply-imputed set, a data analysis system must be able to handle, in addition to the full data matrix, an auxiliary matrix of size: (the number of multiple imputations per missing value) × (the number of missing values). Supposing m imputations per missing value and g percent missing values, this auxiliary data matrix is mg percent as large as the originally hoped-for data matrix. For common values of m and g, this auxiliary matrix is relatively small. For example, suppose that of 30 variables in a survey, 10 are fully observed, 10 are 10% missing, and 10 are 20% missing; then $g = 10$, and for $m = 3$ the auxiliary matrix is only 30% as large as the original data matrix.

Finally, the extra work needed to analyze properly a multiply-imputed data set is undeniable since standard complete-data analyses must be performed on each completed data set and the resultant answers combined within each model represented by the multiple imputations. Often, however, once a complete-data analysis is decided upon, the extra effort to perform it repeatedly is not great and is often measured in computer time rather than investigator time. Combining the resultant answers often only requires the calculation of the means and variances of the repeated complete-data statistics found under each model. The artificial example in Section 1.6 and the real data examples in later chapters illustrate these points too. In cases with large data sets that are expensive to analyze repeatedly, it may make sense to run exploratory analyses using only one or two of the data sets created by multiple imputation, and use all data sets created by the multiple imputations only for final analyses.

The extra effort needed to create, store, and analyze a multiply-imputed data set rather than a singly-imputed data set appears to be quite modest considering the dual payoffs of valid inferences under models and evaluation of sensitivity to models. In many cases, this extra effort seems especially modest when compared with the extra effort needed to work directly with explicit probability models for nonresponse.

1.6. NUMERICAL EXAMPLE USING MULTIPLE IMPUTATION

In order to illustrate the key practical ideas underlying multiple imputation, we present a very small numerical example. This example is not meant to illustrate good survey design, good estimation procedures, or good imputation techniques, but simply the multiple-imputation methodology.

Suppose we have taken a simple random sample of $n = 10$ units from a population of $N = 1000$ units. We know the value of the covariate X (e.g., 1970 size) for each of the N units and try to record the variable Y (e.g., 1980 size) for each of the n units included in the sample, but two units refuse to respond. The objective of the survey is to estimate $\bar{Y}$, the mean of Y in the population. We assume that with complete data, the ratio estimator $\bar{X}\bar{y}/\bar{x}$ would be used with associated approximate 95% interval $\bar{X}\bar{y}/\bar{x} \pm 1.96 SD/n^{1/2}$, where $\bar{X}$ is the known mean of X in the population, say 12, $\bar{y}$ and $\bar{x}$ are the means of Y and X in the random sample of n units, and

$$SD^2 = \sum_{i \in inc} (Y_i - X_i \bar{y}/\bar{x})^2 / (n-1)$$

where *inc* refers to the set of units *inc*luded in the sample. Motivation for the estimator and the associated 95% interval is given in Chapter 2.

Table 1.1a presents the values of Y_i, X_i for the 10 units in the sample, where the question marks indicate missing data; the general notation is $Y_{inc} = (Y_{obs}, Y_{mis})$ where *obs* is the set of indices of *obs*erved values of Y and *mis* is the set of indices of *mis*sing values of Y. Table 1.1b presents four imputations for each of the missing values of Y, two repeated imputations under each of two models. In general, any number of models can be used with any number of repetitions within each model. Shortly, we describe how these imputed values were created, but for now we concentrate on the analysis of the multiply imputed data set given by Tables 1.1a and 1.1b, assuming the imputations in Table 1.1b have been provided to us.

Analyzing This Multiply-Imputed Data Set

For each set of imputations—that is, for each column of Table 1.1b—create a complete data set; these are displayed in Tables 1.1c–1.1f. Then analyze each completed data set as if there had been no nonresponse. Table 1.2a presents the estimates and variances associated with each of the four data sets: the estimators are $\bar{X}\bar{y}/\bar{x}$ and the variances are SD^2/n. Next, combine the two answers obtained under the same model to obtain one inference for $\bar{Y}$ under each model as displayed in Table 1.2b. The center of the resultant interval is the average of the estimates, and the variance associated with this

TABLE 1.1. Artificial example of survey data and multiple imputation.

(a) Observed data

	Y	X
1	10	8
2	?	9
3	14	11
4	?	13
5	16	16
6	15	18
7	20	6
8	4	4
9	18	20
10	22	25

(b) Multiple imputations

	Model 1		Model 2	
	Repetition		Repetition	
	1	2	1	2
Unit 2	10	14	12	17
Unit 4	16	14	19	17

(c) Complete Data Set 1 Model 1, Rep 1

	Y	X
1	10	8
2	10	9
3	14	11
4	16	13
5	16	16
6	15	18
7	20	6
8	4	4
9	18	20
10	22	25
	14.5	13

(d) Complete Data Set 2 Model 1, Rep 2

	Y	X
1	10	8
2	14	9
3	14	11
4	14	13
5	16	16
6	15	18
7	20	6
8	4	4
9	18	20
10	22	25
	14.7	13

(e) Complete Data Set 3 Model 2, Rep 1

	Y	X
1	10	8
2	12	9
3	14	11
4	19	13
5	16	16
6	15	18
7	20	6
8	4	4
9	18	20
10	22	25
	15	13

(f) Complete Data Set 4 Model 2, Rep 2

	Y	X
1	10	8
2	17	9
3	14	11
4	17	13
5	16	16
6	15	18
7	20	6
8	4	4
9	18	20
10	22	25
	15.3	13

TABLE 1.2. Analysis of multiply-imputed data set of Table 1.1.

(a) Ratio Estimates and Associated Variances of Estimation for Each Completed-Data Set

	Model 1		Model 2	
	Repetition		Repetition	
	1	2	1	2
Estimate	13.38	13.57	13.85	14.12
Variance	2.96	3.19	3.38	3.84

(b) Combined Estimates and Variances from Table 1.2a

	Model 1	Model 2
Estimate	13.48	13.99
Variance	3.10	3.66

(c) 95% Intervals for Mean Y in Population Obtained from Table 1.2b

Model 1	Model 2
(10.0, 16.9)	(10.2, 17.7)

estimate has two components: (i) the average within-imputation variance associated with the estimate and (ii) the between-imputation variance of the estimate. Thus, under model 1, the estimate is $(13.38 + 13.57)/2 = 13.48$; the associated estimated average within variance is $(2.96 + 3.19)/2$, and the associated estimated between variance is $[(13.38 - 13.48)^2 + (13.57 - 13.48)^2]$. The estimated variances are combined as (estimated total variance) = (estimated average within variance) + $(1 + m^{-1}) \times$ (estimated between variance), where the factor $(1 + m^{-1})$ multiplying the usual unbiased estimate of between variance is an adjustment for using a finite number of imputations. The associated 95% interval estimate for $\bar{Y}$ is (10.0, 16.9) under model 1 and (10.2, 17.7) under model 2. In practice, better intervals can be formed by calculating degrees of freedom as a simple function of the variance components and using the 95% points appropriate to the corresponding t distribution; when either m is large or the between-variance component is small relative to the total variance (as in this artificial example) the degrees of freedom will be large and thus the normal 95% points will be used. The ratio of the between variance to the total variance estimates the fraction of information missing about $\bar{Y}$ due to nonresponse. This relationship is made precise in Chapter 3.

The essential feature to notice in this illustrative analysis is that only complete-data methods of analysis are needed. We merely have to perform the complete-data analysis that would have been used in the absence of nonresponse on each of the complete data sets created by the multiple imputations. The resultant answers under each model are then easily combined to give one inference under each model.

Creating This Multiply-Imputed Data Set

We now describe how the multiple imputations given in Table 1.1b were generated. Model 1 is an "ignorable" model for nonresponse; ignorable is defined precisely in Chapter 2, but essentially it means that a nonrespondent is just like a respondent with the same value of X. Model 2 is a nonignorable model and posits a systematic difference between respondents and nonrespondents with the same value of X.

The repeated imputations under each model are based on a procedure closely related to the hot-deck. We are not advocating the exact procedure we describe, but rather indicating how a simple intuitive method can be used. More principled methods for creating multiple imputations are presented in Chapters 5 and 6.

For each nonrespondent, we find the two closest matches among the respondents, where by a match we mean having the closest values of X. For the first nonrespondent, unit 2, the two closest matches are units 1 and 3, and for the second nonrespondent, unit 4, the two closest matches are units 3 and 5. The repeated imputations for the missing values Y_{mis} are created by drawing at random from the two closest matches. For the ignorable model, we simply impute the value of Y_i provided by the matching respondent. The first two columns of Table 1.1b give the result. For the nonignorable model, we simply suppose that the nonresponse bias is such that a nonrespondent will tend to have a value of Y_i 20% higher than the matching respondent's value of Y_i. The last two columns of Table 1.1b give the result where the Y_i values have been rounded to the nearest integer.

1.7. GUIDANCE FOR THE READER

Even though multiple imputation is a tool for applied statistics and the intention of this text is to improve the handling of nonresponse in real surveys through the use of multiple imputation, a substantial amount of theoretical development is needed to justify the ideas and procedures such as illustrated in the example of Section 1.6. In particular, a satisfactory development of Bayesian inference in sample surveys in the presence of nonresponse and its relationship to standard randomization-based inference is not a part of the statistical literature, so this is attempted in Chapter 2. This material then forms the conceptual foundation for the Bayesian theory underlying the construction and analysis of a multiply-imputed data set presented in Chapter 3, and the randomization-based evaluation of the resulting procedures presented in Chapter 4. Chapters 5 and 6 primarily deal with the creation of a multiply-imputed data set in ignorable and nonignorable situations, respectively, and although they have less new conceptual material than Chapters 2–4, they still do contain some theory concerning how to draw imputed values under explicit and implicit statistical models that draws on the preceding development. Consequently, even though examples are used throughout to illustrate ideas and tangential mathematical formality is avoided, some special guidance may be useful for the reader anxious to obtain an intuitive appreciation for the straightforward application of multiple imputation before delving into its theoretical underpinnings.

A careful reading of the example of Section 1.6 and the first two sections of Chapter 2 should enable the reader who is familiar with standard statistical methods—both frequentist and Bayesian—to follow the summary description of procedures for creating and analyzing a multiply-imputed data set given in Section 3.1 and the evaluation of these procedures summarized in Section 4.1. The more applied Chapters 5 and 6 can then in large part be read, at least at a relatively superficial level, with occasional reference to specific examples of analyses and procedures from Chapters 2–4. Parts of Chapters 5 and 6, however, especially Sections 5.2 and 6.3, will be difficult to follow before reading at least Sections 2.3–2.6.

As with any statistical procedure, a full appreciation of its application requires an understanding of its conceptual bases, so after obtaining an intuitive appreciation for the use of multiple imputation by following this short-cut path through the text, the applied reader should return to the conceptual chapters 2–4 before attempting to penetrate the subtleties of Chapters 5 and 6.

PROBLEMS

1. Describe the differences between a census and a sample survey. Is a census more immune to problems of nonresponse than a sample survey? Describe a situation in which a well-designed sample survey of a population could be less affected by nonresponse than an equivalently costly census of the same population.
2. Describe a survey you are familiar with that suffers from nonresponse (e.g., a telephone survey for presidential candidates) and discuss possible systematic differences between respondents and nonrespondents and how those differences could bias conclusions from the survey.

3. For three variables, create a pattern of missing and observed values such that the "pairwise-present" correlation matrix, which consists of correlations based on all units responding to both variables in the correlation, is negative definite.

4. Summarize major objectives when handling nonresponse, and advantages and disadvantages of imputation for missing values.

5. For a sample survey with n units and p variables, np data values are intended to be recorded for data analysis. These will be stored in an $n \times p$ data matrix as in Figure 1.1. Suppose there are k missing values and m imputations are used for each missing value. How big a matrix in addition to the original $n \times p$ data matrix is needed to store the multiple imputations? Suppose one analysis would be performed if there had been no missing data. How many complete-data analyses are required of the multiply-imputed data set? If exploratory data analyses are being done initially, should we do all exploratory analyses on all completed data sets or just the final confirmatory analysis?

6. Why is it generally necessary to use multiple rather than single imputation for each missing value? Consider issues of point estimation, interval estimation, and model sensitivity.

7. Summarize some of the literature on response behavior (e.g., references in Section 1.4).

8. Create an example analogous to the one in Section 1.6 where a different method is used for imputation (e.g., a ratio or regression-based method), and produce all steps analogous to those displayed in Tables 1.1 and 1.2.

9. Describe a realistic survey in which there would be few good predictors of missing values and another survey in which there would exist many good predictors of missing values. Comment on the existence of good predictors in the four examples presented here and relate these to the intuitive idea of the fraction of information missing due to nonresponse.

10. Describe factors that create the need for more care when creating imputations and that create the need for larger m.

The following problems all concern the following situation: Consider a simple random sample of size n with n_1 respondents and $n_0 = n - n_1$ nonrespondents, where $\bar{y}_1$ and s_1^2 are the sample mean and variance of the respondents' data, and $\bar{y}_0$ and s_0^2 are the sample mean and variance of the imputed data.

11. Show that the mean and variance of all the data, $\bar{y}_*$ and s_*^2, can be written as

$$\bar{y}_* = (n_1 \bar{y}_1 + n_0 \bar{y}_0)/n$$

and

$$s_*^2 = \left[(n_1 - 1)s_1^2 + (n_0 - 1)s_0^2 + n_1 n_0 (\bar{y}_1 - \bar{y}_0)^2/n\right]/(n-1).$$

12. Suppose imputations are randomly drawn with replacement from the n_1 respondents' values.
   (a) Show that $\bar{y}_*$ is unbiased for the population mean $\bar{Y}$.
   (b) Show that conditional on the observed data, the variance of $\bar{y}_*$ is $n_0 s_1^2 (1 - n_1^{-1})/n^2$ and that the expectation of s_*^2 is $s_1^2 (1 - n_1^{-1}) \times [1 + n_1 n^{-1}(n-1)^{-1}]$.
   (c) Show that conditional on the sample sizes n and n_1 (and the population Y values), the variance of $\bar{y}_*$ is the variance of $\bar{y}_1$ times $[1 + (n_1 - 1)n^{-1}(1 - n_1/n)(1 - n_1/N)^{-1}]$ and show that this is greater than the expectation of $s_*^2(n^{-1} - N^{-1})$.
   (d) Assume n_1 and N/n_1 are large and show that interval estimates of $\bar{Y}$ based on $s_*^2(n^{-1} - N^{-1})$ as the estimated variance of $\bar{y}_*$ are too short by a factor $(1 + nn_1^{-1} - n_1 n^{-1})^{1/2}$. Note that there are two reasons: $n > n_1$ and $\bar{y}_*$ is not as efficient as $\bar{y}_1$. Tabulate true coverages and true significance levels as functions of n_1/n and nominal level.

13. Suppose multiple imputations are created using the method of Problem 12 m times. Let $\bar{\bar{y}}_*$ be the mean of the $\bar{y}_*$ means and T_* be the multiple-imputation estimate of the variance of $\bar{\bar{y}}_*$, $T_* = \bar{U}_* + (1 + m^{-1}) B_*$.
   (a) Show that conditional on the data, the expected value of B_* equals the variance of $\bar{y}_*$.
   (b) Derive the variance of $\bar{\bar{y}}_*$ (conditional on n, n_1, and the population values) and conclude that $\bar{\bar{y}}_*$ with $m > 1$ is more efficient than $\bar{y}_*$.
   (c) Tabulate values of the relative efficiency of $\bar{\bar{y}}_*$ to $\bar{y}_1$ for large n_1 and N/n_1.
   (d) Show that the variance of $\bar{\bar{y}}_*$ (conditional on n, n_1, and the population Y values) is greater than the expectation of T_* by approximately $s_1^2(1 - n_1/n)^2/n_1$.

(e) Assume n_1 and N/n_1 are large and tabulate true coverages and significance levels as functions of n_1/n and nominal level. Compare with the results in Problem 12 part (d).

14. (a) Modify the multiple-imputation procedure of Problem 13 to give the correct answer for large n_1 and N/n_1. Hint: For example, add $s_1 n_1^{-1/2} z_l$ to each imputed value at the lth set of imputations, where z_l is an independent standard normal deviate.

(b) Justify the adjustment in (a) based on the sampling variability of $(\bar{y}_1 - \bar{Y})$.

CHAPTER 2

Statistical Background

2.1. INTRODUCTION

In contrast to much of mathematical statistics, which concentrates on the estimation of hypothetical unknown parameters (e.g., the mean of the hypothetical normal distribution that generated the data), survey methodology concentrates on the estimation of unknown observables such as the average income of families in a particular finite population. Throughout, we use the noun *parameter* to denote only hypothetical unknown and unknowable values. Unobserved values in a finite population are considered to be missing data, that is, missing values of observable quantities. The observable quantities of interest in a finite population will be arrayed in a units(rows)-by-variables(columns) matrix. In any sample survey of the population, only some of the values in this matrix are observed. In a complete census, an attempt is made to observe all values in the matrix.

The approach presented here is an extension of that in Rubin (1978b, 1980a, 1983a) and treats both Bayesian and randomization theory.

Random Indexing of Units

We will let the number of units in the finite population be denoted by N, and throughout we will assume that the indices have been assigned to units as a random permutation of $1, \ldots, N$.

It is important to realize that this assumption is nothing more than a notational convenience used to avoid confusion about the inferential content of indices. Any information in the units' labels or names can always be encoded explicitly as variables. That is, if there exists auxiliary information available on the units and we wish to incorporate it in our survey design or data analysis, then without loss of generality this information can be used to create covariates or stratification variables. For example, suppose the units are people and the names of the people are considered to be informative because of indications of family ties and gender; then covariates can be defined that encode gender and family ties from the names of the people, and the indices can be assigned as a random permutation with no loss of information. Or suppose the units are school districts and the size of the districts are to be used to make sampling decisions as when sampling with probability proportional to size; then size (or a function of size such as stratum indicators "large," "medium," "small") can be a covariate available for each school district, and then again unit indices can be randomly assigned with no loss of information.

2.2. VARIABLES IN THE FINITE POPULATION

Two kinds of variables are defined in the finite population of N units. The first kind includes variables that describe characteristics of units that are of intrinsic interest to the investigator: covariates X and outcome variables Y. The second kind includes indicator variables that are needed to describe which values are observed and which are unobserved and to draw inferences for population quantities: sampling indicators I and response indicators R. Figure 2.1 displays the units by variables matrix in the finite population.

Covariates X

We will let X refer to fully observed covariates, such as stratum indicators or size of unit measurements, recorded for all N units in the population:

$$X = \begin{pmatrix} X_1 \\ \vdots \\ X_N \end{pmatrix}$$

where X_i will be a row vector if there exists more than one fully observed covariate. Often in practical problems, X_i will have many components such as demographic characteristics available on each unit in the population. For example, letting the units be people where $X_i = (X_{i1}, X_{i2}, X_{i3})$, X_{i1} could indicate the gender of the ith unit ($X_{i1} = 1$ for male, $X_{i1} = 0$ for female), X_{i2} could give the medium income in the ZIP code area of the residence of the ith unit (in dollars), and X_{i3} could indicate the education of the ith unit ($X_{i3} = 1$ for high school degree, $X_{i3} = 0$ for no high school degree).

VARIABLES IN THE FINITE POPULATION

	Covariates X		Outcome Variables Y		Sampling Indicators I		Response Indicators R	
	1 $\cdots$ q		1 $\cdots$ p		1 $\cdots$ p		1 $\cdots$ p	
Units in finite population	1	X_{11} $\cdots$ X_{1q}	Y_{11} $\cdots$ Y_{1p}	I_{11} $\cdots$ I_{1p}	R_{11} $\cdots$ R_{1p}
	$\vdots$	$\vdots$	$\vdots$	$\vdots$	$\vdots$
	i	X_{i1} $\cdots$ X_{iq}	Y_{i1} $\cdots$ Y_{ip}	I_{i1} $\cdots$ I_{ip}	R_{i1} $\cdots$ R_{ip}
	$\vdots$	$\vdots$	$\vdots$	$\vdots$	$\vdots$
	N	X_{N1} $\cdots$ X_{Nq}	Y_{N1} $\cdots$ Y_{Np}	I_{N1} $\cdots$ I_{Np}	R_{N1} $\cdots$ R_{Np}

Figure 2.1. Matrix of variables in a finite population of N units.

Outcome Variables Y

We will let Y refer to variables whose values are not known for all units in the population:

$$Y = \begin{pmatrix} Y_1 \\ \vdots \\ Y_N \end{pmatrix}$$

where as with X_i, Y_i will be a row vector if there exists more than one outcome variable. Usually Y will include only those variables of primary interest in the survey, such as individuals' earned income. In cases with background variables that are not fully observed, however, Y will include those variables as well, thereby leaving X fully observed. In order to focus on problems of nonresponse, we will assume that when a value for Y is observed, it is observed without error.

Indicator for Inclusion in the Survey I

The variable

$$I = \begin{pmatrix} I_1 \\ \vdots \\ I_N \end{pmatrix}$$

will be the indicator for inclusion/exclusion from the survey. In the case of just one outcome variable Y, the indicator I is binary, with $I_i = 1$ indicating that the ith unit is included in the survey (i.e., an attempt was made to record Y_i) and $I_i = 0$ indicating that the ith unit is excluded from the survey (i.e., no attempt was made to record Y_i). When Y_i represents more than one variable, say p variables, I_i is also a vector with p components, the jth component, I_{ij}, indicating whether an attempt was made to record Y_{ij} ($I_{ij} = 1$) or not ($I_{ij} = 0$). We assume that I is fully observed; that is, I_{ij} is known for all units and variables. In a census, all components of I equal 1; in a sample survey, some components of I are 0 by design. Thus, I is called the sampling indicator.

In complex surveys with several levels of sampling, I reflects the result of all levels of sampling. For example, suppose the units are high school students in the United States, and that the students in the sample were chosen by first selecting a random sample of schools (the school each student attends being encoded in X), and then selecting a random sample of students within each selected school. Then $I_i = (0, \ldots, 0)$ for all students in schools that are not selected at the first stage of sampling, and $I_i = (0, \ldots, 0)$ for all students in selected schools who were not selected at the second stage of sampling; for students who are in schools selected at the first stage and who are selected at the second stage of sampling, at least some of the components of I_i are not zero.

Indicator for Response in the Survey R

The variable

$$R = \begin{pmatrix} R_1 \\ \vdots \\ R_N \end{pmatrix}$$

will be the indicator for respondent or nonrespondent. In the case of just one outcome variable, R_i is binary with $R_i = 1$ indicating that the ith unit will respond (i.e., Y_i will be observed if an attempt is made to record Y_i—that is, if Y_i is included in the survey) and $R_i = 0$ indicating that the ith unit will not respond (i.e., Y_i will not be observed even if Y_i is included in the survey). When Y_i represents p variables, R_i is also a vector with p components, the jth component R_{ij} indicating response or not on Y_{ij}. We assume that R_{ij} is known whenever $I_{ij} = 1$ and unknown whenever $I_{ij} = 0$. That is, response status is known for all Y_{ij} included in the survey and unknown for all Y_{ij} excluded from the survey. Naturally, R is called the response indicator.

Stable Response

An assumption implicit in the representation summarized in Figure 2.1 is that the values of X, Y, and R for this survey are characteristics of the units in the population, and thus that their values cannot change as a function of which values were included in the survey. That is, (X, Y, R) in Figure 2.1

takes the same value for all possible I. If this were not true, repetitions of (X, Y, R) would have to be used to represent the variety of (X, Y, R) values that could be observed—for example, one set of values when half or more of the units in the population are included in the survey and a second set of values when less than half the units are included. This assumption of "stable response" is like that of "no interference between units" in experimental design (Cox, 1958, p. 19), or more generally, "stable unit-treatment values" in studies for causal effects (Rubin, 1980b). Nonresponse can be modeled without the stable-response assumption, but the job can be substantially more demanding.

Surveys with Stages of Sampling

Even with the assumption of stable response, the representation of values in a survey summarized by Figure 2.1 is not formally adequate to fully describe surveys with several stages of sampling and nonresponse possible at each stage. For example, in two-stage sampling with schools as the clusters and students as the units, at the first stage of sampling, schools are drawn and the school principal may refuse to participate. At the second stage of sampling, students are drawn within the chosen clusters, and Y values are observed for the chosen students who are respondents in schools with respondent principals. Or in some sequential surveys, decisions regarding how many units to sample in the second stage are based on analyses of values observed at the first stage (e.g., the sample variance of outcome variables, the proportion of respondents in the sample).

Although in general such multistage surveys require inclusion and response indicators for each stage of sampling, in many common cases the simple indicators I_{ij} and R_{ij} representing whether an attempt was made to observe Y_{ij} and whether the attempt was successful are adequate to summarize essential features of the survey. Multiple imputation is applicable to surveys where this simple notation is not adequate, but the general notation is more cumbersome with no accompanying increase in the clarity of presentation of essential ideas concerning nonresponse. Hence, we restrict our attention to situations where it is adequate to have one indicator for inclusion in the survey and one indicator for response.

2.3. PROBABILITY DISTRIBUTIONS AND RELATED CALCULATIONS

We let $\Pr(\cdot)$ indicate probability or probability density depending on context. Thus $\Pr(I)$ is the probability of a particular value of the sampling indicator I. With one outcome variable, so that I_i is binary, and a simple random sample of size n, we have

$$\Pr(I) = 1 \Big/ \binom{N}{n} \text{ if } \sum_{1}^{N} I_i = n \text{ and } \Pr(I) = 0 \text{ if } \sum_{1}^{N} I_i \neq n. \quad (2.3.1)$$

With one outcome variable whose N components are independent and identically distributed (i.i.d.) according to the standard normal distribution [i.e., the normal distribution with mean zero and variance one, $N(0,1)$], we have

$$\Pr(Y) = \prod_{i=1}^{N} (2\pi)^{-1/2} \exp(-Y_i^2/2). \quad (2.3.2)$$

Conditional Probability Distributions

We let $\Pr(\cdot|\cdot)$ refer to conditional probability or probability density, depending on context. Thus $\Pr(I|X)$ is the probability of a particular value of the sampling indicator I given a particular value of the covariate X. Suppose we have only one outcome variable and only one covariate indicating size of the unit; further suppose each unit is independently chosen to be included in the sample with probability proportional to size, say with probability X_i/X_{max} where X_{max} is the largest X_i. Then

$$\Pr(I|X) = \prod_{i=1}^{N} (X_i/X_{max})^{I_i} [1 - (X_i/X_{max})]^{1-I_i}. \quad (2.3.3)$$

For a second example of the notation, suppose that given the one covariate X, the N components of the outcome variable Y are independent normals with Y_i having mean X_i and variance 1. Then

$$\Pr(Y|X) = \prod_{i=1}^{N} (2\pi)^{-1/2} \exp[-(Y_i - X_i)^2/2]. \quad (2.3.4)$$

Probability Specifications Are Symmetric in Unit Indices

As a reflection or formalization of the assumption of random indexing of units, all probability specifications will be symmetric (or exchangeable) in the unit indices. That is, $\Pr(A_1, \ldots, A_N) = \Pr(A_{p_1}, \ldots, A_{p_N})$ where $p_1, \ldots, p_N$ is a permutation of $1, \ldots, N$. The previous examples of probability specifications for $\Pr(I)$ in (2.3.1), $\Pr(Y)$ in (2.3.2), $\Pr(I|X)$ in (2.3.3) and $\Pr(Y|X)$ in (2.3.4) all satisfied this condition. The following example does not:

$$\Pr(I) = \begin{cases} 1 & \text{if } I_i = 1 \text{ for all even } i \text{ and } I_i = 0 \text{ for all odd } i \\ 0 & \text{otherwise.} \end{cases}$$

Bayes's Theorem

Bayes's theorem is a standard tool of mathematical statistics. It is used to relate conditional distributions and marginal distributions. If A and B are two random variables with joint distribution $\Pr(A, B)$, then

$$\Pr(A, B) = \Pr(A|B)\Pr(B) = \Pr(A)\Pr(B|A). \qquad (2.3.5)$$

Bayes's theorem is usually stated as

$$\Pr(A|B) = \Pr(A)\Pr(B|A)/\Pr(B). \qquad (2.3.6)$$

If there exists an additional variable, C, being conditioned upon

$$\Pr(A, B|C) = \Pr(A|B, C)\Pr(B|C) = \Pr(A|C)\Pr(B|A, C) \quad (2.3.7)$$

and

$$\Pr(A|B, C) = \Pr(A|C)\Pr(B|A, C)/\Pr(B|C).$$

Of course, marginal and joint distributions are related by

$$\Pr(B) = \int \Pr(A, B) \, dA \qquad (2.3.8)$$

and

$$\Pr(B|C) = \int \Pr(A, B|C) \, dA \qquad (2.3.9)$$

where $\int [\cdot] \, dA$ represents integration or summation over all possible values of A depending on whether A is continuous or discrete. Also, $\int \Pr(A) \, dA = 1$ and $\int \Pr(A|B) \, dA = 1$ for all B since distributions must sum to 1.

For discussion of such relations and the use of Bayes's theorem in applied inference, see, for example, Box and Tiao (1973, Section 1.2).

Finding Means and Variances from Conditional Means and Variances

Let $E(A)$ represent the expectation (mean) of the scalar random variable A,

$$E(A) = \int A \Pr(A) \, dA, \qquad (2.3.10)$$

and $V(A)$ represent the variance of A

$$V(A) = \int A^2 \Pr(A) \, dA - E(A)^2. \qquad (2.3.11)$$

Also, let $E(A|B)$ represent the conditional expectation of A given a fixed value of B (a function of the value of B),

$$E(A|B) = \int A \Pr(A|B) \, dA, \qquad (2.3.12)$$

and $V(A|B)$ represent the conditional variance of A given a fixed value of B (also a function of the value of B),

$$V(A|B) = \int A^2 \Pr(A|B) \, dA - E(A|B)^2. \qquad (2.3.13)$$

The notations $E(A|B)$ and $V(A|B)$ will always imply integration over all random variables except B. For example, if we introduce another random variable C,

$$E(A|B) = \iint A \Pr(A, C|B) \, dA \, dC$$

and

$$E(A) = \iiint A \Pr(A, B, C) \, dA \, dB \, dC.$$

Equations (2.3.10), (2.3.12), and Bayes's theorem can be used to prove that the expectation of A equals the expectation, over the distribution of B, of the conditional expectation of A given B:

$$E(A) = E[E(A|B)]. \qquad (2.3.14)$$

That is, by (2.3.10) and (2.3.12), the assertion is that

$$\int A \Pr(A) \, dA = \int \left[\int A \Pr(A|B) \, dA \right] \Pr(B) \, dB. \qquad (2.3.15)$$

But the right-hand side of (2.3.15) is

$$\iint A \Pr(A|B) \Pr(B) \, dA \, dB,$$

which equals, by Bayes's theorem,

$$\iint A \Pr(A) \Pr(B|A) \, dA \, dB,$$

which equals

$$\int A \Pr(A) \left[\int \Pr(B|A) \, dB \right] dA,$$

where the factor in brackets is one for all A.

Similarly, equations (2.3.11), (2.3.13), and Bayes's theorem can be used to prove that the variance of A equals the sum of (a) the variance (over B) of the conditional expectation of A given B and (b) the expectation (over B) of the conditional variance of A given B:

$$V(A) = V[E(A|B)] + E[V(A|B)]. \qquad (2.3.16)$$

Equations (2.3.14) and (2.3.16) also hold when A is a vector random variable. Then $E(\cdot)$ is a vector of the same dimension and $V(\cdot)$ is a square variance–covariance matrix. Also, the quantities $(\cdot)^2$ in equations (2.3.11) and (2.3.13) must be replaced by the outer product of the vectors, $(\cdot)'(\cdot)$, where $(\cdot)$ is a row vector.

Equations (2.3.14) and (2.3.16) are very useful in many calculations, and we use them throughout this text.

2.4. PROBABILITY SPECIFICATIONS FOR INDICATOR VARIABLES

Probability specifications for the indicator variables I and R are needed to draw inferences for unknown values in the finite population. We call the models for these specifications "mechanisms" rather than "models" to distinguish them from the more usual statistical models relating to the distribution of the variables X, Y.

Sampling Mechanisms

The specification for $\Pr(I|X, Y, R)$ is called the sampling mechanism. All formal frameworks currently used to draw inferences in surveys require an explicit or implicit description of the sampling mechanism.

The sampling mechanism is said to be *unconfounded*, or more precisely unconfounded with (Y, R) if

$$\Pr(I|X, Y, R) = \Pr(I|X) \quad \text{for all possible } (X, Y, R, I). \quad (2.4.1)$$

If $\Pr(I|X, Y, R)$ depends on (Y, R), then the sampling mechanism is confounded[†] with (Y, R). The sampling mechanism is unconfounded with R if

$$\Pr(I|X, Y, R) = \Pr(I|X, Y), \quad (2.4.2)$$

and unconfounded with Y if

$$\Pr(I|X, Y, R) = \Pr(I|X, R). \quad (2.4.3)$$

A sampling mechanism is said to be *probability* if it assigns positive probability to each Y_{ij} being included in the sample:

$$\Pr(I_{ij} = 1|X, Y, R) > 0 \quad \text{for all } I_{ij}. \quad (2.4.4)$$

If $\Pr(I_{ij} = 1|X, Y, R) = 0$ for some unit i and some Y variable j, then the sampling mechanism is not a probability sampling mechanism.

Standard scientific sampling techniques are designed so that there will be one commonly accepted specification for $\Pr(I|X, Y, R)$. Often, it will be an unconfounded probability sampling mechanism, as defined by (2.4.1) and (2.4.4). Essentially, these simply imply first that the sampling mechanism can use observed X variables such as stratum indicators to choose samples but cannot use values of Y or R, and second that each value of Y in the population has a chance of being included in the sample. Thus, this definition of unconfounded probability sampling mechanisms includes standard scientific methods of sampling such as stratified random sampling, cluster sampling, and probability-proportional-to-size sampling but excludes sequential sampling methods where sampling decisions depend on values of previously observed Y_{ij} or R_{ij}. A definition of scientific sampling mechanisms appropriate to complex multistage or sequential surveys would allow dependence on Y, R, or I values observed from a previous stage; to be precise, such a definition would need notation for inclusion and response at each stage.

[†]When the quantity being conditioned upon is a random variable, "unconfounded" is the same as "independent"; recently Dawid (1979) has used "independent" even when the quantity being conditioned upon is not a random variable, but statistical tradition (i.e., in experimental design) has avoided "independent" in favor of "unconfounded" in such cases.

Examples of Unconfounded Probability Sampling Mechanisms

A specific example of an unconfounded probability sampling mechanism is given by simple random sampling for scalar Y_i, where $\Pr(I|X, Y, R) = \Pr(I)$ is given by equation (2.3.1). Another example of a simple unconfounded probability sampling mechanism that does not involve a covariate is Bernoulli sampling for scalar Y_i with probability .1 of being included in the sample:

$$\Pr(I|X, Y, R) = \Pr(I) = \prod_{i=1}^{N} (.1)^{I_i}(.9)^{1-I_i}. \qquad (2.4.5)$$

A specific example of an unconfounded probability sampling mechanism involving a covariate is given by (2.3.3). For a similar example with scalar Y_i, suppose $X_i > 0$ is a scalar measure of size of the ith unit, where larger units are more likely to be included according to

$$\Pr(I|Y, X, R) = \Pr(I|X) = \prod_{i=1}^{N} \left(\frac{X_i}{1 + X_i}\right)^{I_i} \left(\frac{1}{1 + X_i}\right)^{1-I_i}. \qquad (2.4.6)$$

Note that if size had not been observed for all units, size would have been a Y variable, not an X variable, and the resulting sampling with probability involving size would have been confounded. The important point is that either the values of variables used to make decisions about which units should be included in the sample—or summaries of these variables from which the probability of I can be calculated—must be recorded for the sampling mechanism to be unconfounded.

The sampling mechanisms given by (2.3.1), (2.3.3), (2.4.5), and (2.4.6) are for scalar Y_i. Commonly in practice, however, I is such that $I_i = (I_{i1}, \ldots, I_{ip})$ is either all ones or all zeroes, so that if a unit is included in the sample, all variables are included and if the unit is excluded, all variables are excluded. In this case with unconfounded sampling mechanisms, the distinction between Y_i with one and many components is irrelevant to the definition of the sampling mechanism, and so (2.3.1), (2.3.3), (2.4.5), and (2.4.6) are examples of unconfounded probability sampling mechanisms for p-variate Y_i (with the interpretation that I_i in the expression equals $I_{i1} = \cdots = I_{ip}$). With confounded sampling mechanisms and $I_i = (1, \ldots, 1)$ or $(0, \ldots, 0)$, the distinction between scalar and p-variate Y_i is still relevant because the probability of selection could depend, for example, on Y_{i1} but not on Y_{i2}.

Examples of Confounded and Nonprobability Sampling Mechanisms

In nonscientific surveys, the probability of being included in the sample might reasonably be considered to be correlated with Y_i, for example, the more active sportsman may be more likely to be given a questionnaire asking the proportion of each day devoted to sporting activities (Y_i). For such a situation, Bernoulli sampling with probability of inclusion increasing with Y_i might be plausible, for example,

$$\Pr(I|X, Y, R) = \prod_{i=1}^{N} \left(\frac{Y_i}{1 + Y_i}\right)^{I_i} \left(\frac{1}{1 + Y_i}\right)^{1-I_i}. \quad (2.4.7)$$

This sampling mechanism is confounded with Y, but is a probability sampling mechanism if no Y_i is zero.

For an example of a sampling mechanism confounded with R, consider a voluntary mail survey of an entire population, where the returned questionnaires are considered to constitute the sample survey; since respondents are included in the survey and nonrespondents are excluded,

$$\Pr(I|X, Y, R) = \begin{cases} 1 & \text{if } I = R, \\ 0 & \text{otherwise.} \end{cases} \quad (2.4.8)$$

For an example of an unconfounded nonprobability sampling mechanism, suppose a survey is taken by telephone in a population that includes some homes without telephones. If X_i is the number of telephones in the home and a census is attempted, then we have

$$\Pr(I|X, Y, R) = \begin{cases} 1 & \text{if } I_i = 1 \text{ when } X_i > 0, \text{ and } I_i = 0 \text{ when } X_i = 0, \\ 0 & \text{otherwise.} \end{cases}$$
$$(2.4.9)$$

Response Mechanisms

In order to draw inferences when there is the possibility of nonresponse, standard frameworks, either explicitly or implicitly, require a specification $\Pr(R|X, Y)$. Because both R and I are $N \times p$ 0–1 indicator random variables, examples of specifications for $\Pr(I|X, Y)$ are examples of specifications for $\Pr(R|X, Y)$ with the notational replacement of R for I. Most of our examples of sampling mechanisms were unconfounded with R and thus satisified

$$\Pr(I|X, Y, R) = \Pr(I|X, Y).$$

PROBABILITY SPECIFICATIONS FOR (X, Y) 39

Consequently, with R_i in place of I_i, the right-hand sides of (2.3.1), (2.3.3), (2.4.5), (2.4.6), (2.4.7), and (2.4.9) all provide examples of response mechanisms; for example, in (2.4.9), Y_i could be income in $100,000 increments, reflecting an increasing nonresponse rate with increasing income that might be plausible for describing nonresponse in an income survey.

Furthermore, definitions analogous to those for unconfounded and probability sampling mechanisms can be made for response mechanisms. If

$$\Pr(R|X, Y) = \Pr(R|X), \qquad (2.4.10)$$

the response mechanism is said to be unconfounded. If

$$\Pr(R_{ij}|X, Y) > 0 \quad \text{for all } i, j, \qquad (2.4.11)$$

then the response mechanism is a probability response mechanism. With the notational replacement of R for I, (2.3.1), (2.3.3), (2.4.5), and (2.4.6) all provide examples of unconfounded probability response mechanisms if $X_i > 0$. An unconfounded probability response mechanism is just like another level of unconfounded probability sampling of the values included in the sample, where the resultant sampling indicators are $I_{ij}R_{ij}$. Usually, nonresponse cannot be realistically assumed to be unconfounded, and then assumptions about specific forms for the response mechanism can be crucial to appropriate adjustments for nonresponse.

2.5. PROBABILITY SPECIFICATIONS FOR (X, Y)

Thus far, we have discussed probability specifications for the sampling and response mechanisms. Together, these provide a specification for the joint conditional distribution of I and R given X and Y:

$$\Pr(I, R|X, Y) = \Pr(I|X, Y, R)\Pr(R|X, Y).$$

A natural question that arises is whether to complete the specification of distributions for the variables by adding a specification for the distribution of (X, Y). With randomization theory, the answer is no; with Bayesian theory, the answer is yes, and so we proceed to discuss $\Pr(X, Y)$ and present illustrative Bayesian calculations using particular specifications. The full problem of Bayesian inference in a survey will be introduced in Section 2.6.

de Finetti's Theorem

Because (X, Y) is an $N \times (q + p)$ matrix of random variables, it might seem difficult to formulate a sensible distribution for it. This effort is greatly simplified by appealing to de Finetti's theorem and its extensions (Feller, 1966, p. 225; Hewitt and Savage, 1956; Diaconis, 1977).

Suppose that we have specified a distribution for (X, Y), which by assumption of random assignment of unit indices, must be exchangeable in the unit (row) indices: $\Pr(X, Y) = \Pr(\text{row-perm}(X, Y))$ where row-perm(X, Y) is any permutation of the rows of (X, Y). Then de Finetti's theorem implies that $\Pr(X, Y)$ can be written in a form where, given a vector parameter θ with marginal (prior) density $\Pr(\theta)$, the (X_i, Y_i), $i = 1, \ldots, N$, are independently and identically distributed (i.i.d.) with common distribution $f_{XY}(X_i, Y_i | \theta)$:

$$\Pr(X, Y) = \int \left[\prod_{i=1}^{N} f_{XY}(X_i, Y_i | \theta) \right] \Pr(\theta) \, d\theta. \qquad (2.5.1)$$

Equation (2.5.1) in fact might not hold exactly for some $\Pr(X, Y)$ with continuous (X, Y) and finite N (e.g., see Problem 25), but this makes little difference to practical model-building efforts. The result is of general importance because it means that standard tools of mathematical statistics employing i.i.d. models can be used to draw inferences with essentially no loss of generality.

Some Intuition

This conclusion on exchangeability and i.i.d. models is really intuitively somewhat obvious. Suppose we were faced with N observations that we knew arose from an exchangeable distribution, and we were asked to estimate the dependence between the observations or to decide if they arose from an i.i.d. model. Even when all the observations are nearly the same, how are we to distinguish between (a) an i.i.d. model with very small variance and (b) an exchangeable model with large variance but high correlation? Since we really can never rule out the i.i.d. model except under special cases with rigid assumptions, we may as well assume the simple i.i.d. structure and simplify our modeling efforts.

Example 2.1. A Simple Normal Model for Y_i

As a specific example of the use of i.i.d. modeling, suppose that Y_i is scalar and there is no X. Also, suppose that for $i = 1, \ldots, N$, the distribution of Y_i

PROBABILITY SPECIFICATIONS FOR (X, Y)

given $\theta = (\mu, \sigma^2)$ is i.i.d. $N(\mu, \sigma^2)$, and that the prior distribution of (μ, σ^2) has density proportional to σ^{-2}. Then the distribution of $\bar{Y} = \sum_1^N Y_i/N$ given $Y_1, \ldots, Y_n$ is a t on $n - 1$ degrees of freedom with location $\bar{y}$ and scale $s(n^{-1} - N^{-1})^{1/2}$, where $\bar{y} = \sum_1^n Y_i/n$ and $s^2 = \sum_1^n (y_i - \bar{y})^2/(n-1)$.

The proof is straightforward. Because $\bar{Y} = (\sum_1^n Y_i + \sum_{n+1}^N Y_i)/N$ and $\sum_1^n Y_i$ is known, the critical calculation is to find the conditional distribution of $\sum_{n+1}^N Y_i$ given $Y_1, \ldots, Y_n$. Since given $(\mu, \sigma^2, Y_1, \ldots, Y_n)$, the $Y_{n+1}, \ldots, Y_N$ are i.i.d. $N(\mu, \sigma^2)$, the distribution of $\sum_{n+1}^N Y_i$ given $(\mu, \sigma^2, Y_1, \ldots, Y_n)$ is $N((N-n)\mu, (N-n)\sigma^2)$. By Lemma 2.1 below, the distribution of μ given $(\sigma^2, Y_1, \ldots, Y_n)$ is $N(\bar{y}, \sigma^2/n)$. Hence, given $(\sigma^2, Y_1, \ldots, Y_n)$, $\sum_{n+1}^N Y_i$ is normal with mean $E[(N-n)\mu|\sigma^2, Y_1, \ldots, Y_n] = (N-n)\bar{y}$ and variance $V[(N-n)\mu|\sigma^2, Y_1, \ldots, Y_n] + E[(N-n)\sigma^2|\sigma^2, Y_1, \ldots, Y_n] = (N-n)^2(\sigma^2/n) + (N-n)\sigma^2 = N(N-n)(\sigma^2/n)$. Consequently, given $(\sigma^2, Y_1, \ldots, Y_n)$ the distribution of $\bar{Y} = (n\bar{y} + \sum_{n+1}^N Y_i)/N$ is normal with mean $[n\bar{y} + (N-n)\bar{y}]/N = \bar{y}$ and variance $N^{-2}[N(N-n)\sigma^2/n] = (1 - n/N)\sigma^2/n$. Since by Lemma 2.1 the distribution of σ^2 given $Y_1, \ldots, Y_n$ is $(n-1)s^2\chi^{-2}_{n-1}$, where χ^{-2}_{n-1} is the the inverted χ^2 distribution on $n - 1$ degrees of freedom, it follows that $\bar{Y}$ is t on $n - 1$ degrees of freedom with location $\bar{y}$ and scale $[s^2(n^{-1} - N^{-1})]^{1/2}$.

Lemma 2.1. Distributions Relevant to Example 2.1

Suppose that given (μ, σ^2), $Y_i, i = 1, \ldots, N$ are i.i.d. $N(\mu, \sigma^2)$ and that the marginal (prior) distribution of (μ, σ^2) has density proportional to σ^{-2}; that is, suppose

$$\Pr(Y_1, \ldots, Y_N | \mu, \sigma^2) = \prod_{i=1}^N (2\pi\sigma^2)^{-1/2} \exp\left[-(Y_i - \mu)^2/(2\sigma^2)\right] \quad (2.5.2)$$

and

$$\Pr(\mu, \sigma^2) \propto \sigma^{-2}. \quad (2.5.3)$$

Then (i) the distribution of μ given $(\sigma^2, Y_1, \ldots, Y_n)$ is $N(\bar{y}, \sigma^2/n)$, where $\bar{y} = \sum_1^n Y_i/n$; that is,

$$\Pr(\mu | \sigma^2, Y_1, \ldots, Y_n) = (2\pi\sigma^2/n)^{-1/2} \exp\left[-(\mu - \bar{y})^2/(2\sigma^2/n)\right]; \quad (2.5.4)$$

(ii) the distribution of σ^2 given $(Y_1, \ldots, Y_n)$ is $(n-1)s^2\chi^{-2}_{n-1}$ where $s^2 = \sum_1^n (Y_i - \bar{y})^2/(n-1)$ and χ^{-2}_{n-1} is the inverted chi-squared distribution with

$n-1$ degrees of freedom; that is,

$$\Pr(\sigma^2|Y_1,\ldots,Y_n) = K(Y_1,\ldots,Y_n)(\sigma^2)^{-(n+1)/2}\exp[-(n-1)s^2/(2\sigma^2)],$$

$$\sigma^2 > 0 \quad (2.5.5)$$

where

$$K(Y_1,\ldots,Y_n) = [(n-1)s^2/2]^{n+1/2}/\Gamma[(n-1)/2].$$

The proof is straightforward. From (2.5.2),

$$\Pr(Y_1,\ldots,Y_n|\mu,\sigma^2) = \prod_{i=1}^{n}(2\pi\sigma^2)^{-1/2}\exp[-(Y_i-\mu)^2/(2\sigma^2)]. \quad (2.5.6)$$

Since $\Pr(\mu,\sigma^2|Y_1,\ldots,Y_n) \propto \Pr(Y_1,\ldots,Y_n|\mu,\sigma^2)\Pr(\mu,\sigma^2)$, (2.5.6) and (2.5.3) imply that

$$\Pr(\mu,\sigma^2|Y_1,\ldots,Y_n)$$

$$\propto (\sigma^2)^{-(1+n/2)}\exp\left[-\sum_{1}^{n}(Y_i-\mu)^2/(2\sigma^2)\right] \quad (2.5.7)$$

$$\propto (\sigma^2)^{-(1+n/2)}\exp\left[-n(\bar{y}-\mu)^2/(2\sigma^2)-\sum_{1}^{n}(Y_i-\bar{y})^2/(2\sigma^2)\right]$$

$$\propto (\sigma^2/n)^{-1/2}\exp[-(\bar{y}-\mu)^2/(2\sigma^2/n)]$$

$$\times(\sigma^2)^{-(1+n/2)}\exp[-(n-1)s^2/(2\sigma^2)]. \quad (2.5.8)$$

Expression (2.5.8) proves parts (i) and (ii) because the first factor is proportional to (2.5.4) and the second factor is proportional to (2.5.5).

The reader unfamiliar with these calculations is referred to Box and Tiao (1973, Chapter 2, Section 4) or more specifically, Ericson (1969, p. 205).

Example 2.2. A Generalization of Example 2.1

Again suppose that Y_i is scalar and that there is no X, but drop the assumption of normality. That is, suppose that given θ, the $Y_1,\ldots,Y_N$ are i.i.d. where θ has prior distribution $p(\theta)$. Let $E(Y_i|\theta) = \mu$ and $V(Y_i|\theta) = \sigma^2$, both finite functions of θ. Then the conditional expectation of $\bar{Y}$ given

$Y_1, \ldots, Y_n$ is
$$(n/N)\bar{y} + (1 - n/N)\hat{\mu}$$
where $\hat{\mu} = E(\mu|Y_1, \ldots, Y_n)$, and the conditional variance of $\bar{Y}$ given $Y_1, \ldots, Y_n$ is
$$(1 - n/N)[\hat{\sigma}^2/N + (1 - n/N)V(\mu|Y_1, \ldots, Y_n)]$$
where $\hat{\sigma}^2 = E(\sigma^2|Y_1, \ldots, Y_n)$.

The proof of this claim is straightforward. First,
$$E(\bar{Y}|Y_1, \ldots, Y_n) = E\left(\sum_1^n Y_i + \sum_{n+1}^N Y_i \bigg| Y_1, \ldots, Y_n\right)/N.$$

By definition
$$E\left(\sum_1^n Y_i \bigg| Y_1, \ldots, Y_n\right) = n\bar{y},$$
and by conditioning on θ we can write
$$E\left(\sum_{n+1}^N Y_i \bigg| Y_1, \ldots, Y_n\right) = E\left[E\left(\sum_{n+1}^N Y_i \bigg| Y_1, \ldots, Y_n, \theta\right) \bigg| Y_1, \ldots, Y_n\right]$$
where the outer expectation is over θ and the inner expectation equals $(N - n)\mu$ since given θ, $Y_{n+1}, \ldots, Y_N$ are i.i.d. with mean μ and are independent of $Y_1, \ldots, Y_n$.

Thus
$$E(\bar{Y}|Y_1, \ldots, Y_n) = [n\bar{y} + (N - n)E(\mu|Y_1, \ldots, Y_n)]/N.$$

Next
$$V(\bar{Y}|Y_1, \ldots, Y_n) = V\left(\sum_{n+1}^N Y_i \bigg| Y_1, \ldots, Y_n\right)/N^2.$$

By conditioning on θ we have
$$V\left(\sum_{n+1}^N Y_i \bigg| Y_1, \ldots, Y_n\right) = E\left[V\left(\sum_{n+1}^N Y_i \bigg| Y_1, \ldots, Y_n, \theta\right) \bigg| Y_1, \ldots, Y_n\right]$$
$$+ V\left[E\left(\sum_{n+1}^N Y_i \bigg| Y_1, \ldots, Y_n, \theta\right) \bigg| Y_1, \ldots, Y_n\right].$$

But given θ, $Y_{n+1}, \ldots, Y_N$ are i.i.d. with mean μ and variance σ^2 and are independent of $Y_1, \ldots, Y_n$. Thus

$$V\left(\sum_{n+1}^{N} Y_i \bigg| Y_1, \ldots, Y_n\right) = E\left[(N-n)\sigma^2 | Y_1, \ldots, Y_n\right]$$

$$+ V\left[(N-n)\mu | Y_1, \ldots, Y_n\right],$$

and so

$$V(\bar{Y}|Y_1, \ldots, Y_n) = \left[(N-n)E(\sigma^2|Y_1, \ldots, Y_n)\right.$$

$$\left. + (N-n)^2 V(\mu|Y_1, \ldots, Y_n)\right]/N^2.$$

Example 2.3. An Application of Example 2.2: The Bayesian Bootstrap

Consider the following specification discussed in Ericson (1969) and Rubin (1981a). Let $d = (d_1, \ldots, d_K)$ be the vector of all possible distinct values of Y_i, and let $\theta = (\theta_1, \ldots, \theta_K)$ be the associated vector of probabilities, $\Sigma \theta_i = 1$, where we suppose that $Y_1, \ldots, Y_N$ given θ are i.i.d.:

$$\Pr(Y_i = d_k | \theta) = \theta_k.$$

Suppose further the prior distribution on θ is improper:

$$\Pr(\theta) = \prod_{k=1}^{K} \theta_k^{-1} \text{ if } \Sigma \theta_k = 1, \text{ and } 0 \text{ otherwise}.$$

We will call this the "Bayesian bootstrap" specification, following Rubin (1981a).

Let n_k = the number of Y_i, $i = 1, \ldots, n$, equal to d_k, $\Sigma_1^K n_k = n$. The conditional distribution of θ given $Y_1, \ldots, Y_n$, $\Pr(\theta|Y_1, \ldots, Y_n)$, is then the $(K-1)$-variate Dirichlet distribution proportional to

$$\begin{cases} \prod_{k=1}^{K} \theta_k^{n_k - 1} & \text{if } \Sigma \theta_k = 1 \text{ and } \Sigma n_k = n, \\ 0 & \text{otherwise}. \end{cases}$$

Under this specification, values d_k that are not observed have zero probability given $(Y_1, \ldots, Y_n)$.

Let us apply the result in Example 2.2 in order to calculate the conditional mean and variance of $\bar{Y}$ given $(Y_1, \ldots, Y_n)$ under this model, which some would label "nonparametric."

PROBABILITY SPECIFICATIONS FOR (X, Y)

The mean of Y_i given θ is

$$\mu = \sum_{k=1}^{K} d_k \theta_k$$

and the variance of Y_i given θ is

$$\sigma^2 = \sum_{k=1}^{K} d_k^2 \theta_k - \mu^2.$$

We need to find $\hat{\mu}$ and $\hat{\sigma}^2$, the conditional means of μ and σ^2 given $(Y_1, \ldots, Y_n)$, as well as the conditional variance of μ given $(Y_1, \ldots, Y_n)$. From standard results on the Dirichlet distribution (Wilks, 1963, p. 238), given $(Y_1, \ldots, Y_n)$, the mean of θ is $\hat{\theta} = (n_1, \ldots, n_K)/n$, the variance of θ_k is $\hat{\theta}_k(1 - \hat{\theta}_k)/(n+1)$, and the covariance of θ_k and $\theta_{k'}$ is $-\hat{\theta}_k \hat{\theta}_{k'}/(n+1)$. Thus $\hat{\mu} = \sum d_k \hat{\theta}_k = \sum d_k n_k / n = \bar{y}$, so that the conditional mean of $\bar{Y}$ given $(Y_1, \ldots, Y_n)$ is $\bar{y}$ just as with the normal specification of Example 2.1. Also

$$\hat{\sigma}^2 = \sum d_k^2 \hat{\theta}_k - \left[\bar{y}^2 + V(\mu | Y_1, \ldots, Y_n) \right]$$

$$= \sum_{1}^{n} Y_i^2 / n - \bar{y}^2 - V(\mu | Y_1, \ldots, Y_n)$$

$$= s^2 (1 - n^{-1}) - V(\mu | Y_1, \ldots, Y_n).$$

Thus, the conditional variance of $\bar{Y}$ given $(Y_1, \ldots, Y_n)$ is

$$(1 - n/N)\left[(n-1)s^2/(nN) + V(\mu | Y_1, \ldots, Y_n)(N - n - 1)/N\right].$$

But

$$V(\mu | Y_1, \ldots, Y_n) = \sum d_k^2 \hat{\theta}_k (1 - \hat{\theta}_k)/(n+1) - 2 \sum_{k > k'} \hat{\theta}_k \hat{\theta}_{k'} d_k d_{k'} / (n+1)$$

$$= \sum d_k^2 \hat{\theta}_k / (n+1) - \left[\sum d_k^2 \hat{\theta}_k^2 + 2 \sum_{k > k'} d_k d_{k'} \hat{\theta}_k \hat{\theta}_{k'} \right] / (n+1)$$

$$= \sum d_k^2 \hat{\theta}_k / (n+1) - \left(\sum d_k \hat{\theta}_k \right)^2 / (n+1)$$

$$= s^2 (1 - n^{-1}) / (n+1).$$

Hence, from Example 2.2 and some algebraic manipulation, the conditional variance of $\bar{Y}$ given $(Y_1,\ldots,Y_n)$ is $s^2(n^{-1} - N^{-1})(n-1)/(n+1)$, which is very nearly the same answer as with the normal Y specification of Example 2.1.

Notice that whenever $E(\mu|Y_1,\ldots,Y_n) \doteq \bar{y}$, $V(\mu|Y_1,\ldots,Y_n) \doteq s^2/n$, and $E(\sigma^2|Y_1,\ldots,Y_n) \doteq s^2$ as hold in Example 2.3 for modestly large n, the result in Example 2.2 implies that the conditional mean and variance of $\bar{Y}$ given $(Y_1,\ldots,Y_n)$ will be nearly the same as under the normal specification of Example 2.1. Pratt (1965), for example, argues that these approximations hold rather generally.

Example 2.4. Y_i Approximately Proportional to X_i

Suppose Y_i and X_i are both scalar and (X_i, Y_i) are i.i.d. given the parameter θ. Also suppose that $f_{XY}(X_i, Y_i|\theta)$ is factored into $f_{Y|X}(Y_i|X_i, \theta_{Y|X}) f_X(X_i|\theta_X)$ where $\theta_{Y|X}$ and θ_X are *a priori* independent functions of θ; Y_i given X_i and $\theta_{Y|X}$ is $N(\beta X_i, (\sigma X_i^g)^2)$ where the exponent g is known, and the prior distribution of $\theta_{Y|X} = (\beta, \log \sigma)$ has density proportional to a constant [i.e., $\Pr(\beta, \sigma^2) \propto \sigma^{-2}$]. Then the conditional distribution of $\bar{Y}$ given X and $Y_1,\ldots,Y_n$ is t on $n-1$ degrees of freedom with location

$$\left[n\bar{y} + \hat{\beta} \sum_{n+1}^{N} X_i\right]/N \qquad (2.5.9)$$

and scale

$$\hat{\sigma}\left[\sum_{n+1}^{N} X_i^{2g} + \left(\sum_{n+1}^{N} X_i\right)^2 / \sum_{1}^{n} X_i^{2-2g}\right]^{1/2}/N \qquad (2.5.10)$$

where

$$\hat{\beta} = \sum_{1}^{n} Y_i X_i^{1-2g} / \sum_{1}^{n} X_i^{2-2g} \qquad (2.5.11)$$

and

$$\hat{\sigma}^2 = \sum_{1}^{n} \left[(Y_i - X_i\hat{\beta})/X_i^g\right]^2/(n-1). \qquad (2.5.12)$$

PROBABILITY SPECIFICATIONS FOR (X, Y)

To prove these results, we first find the distribution of $N\overline{Y}$ given $(Y_1, \ldots, Y_n)$, X and θ:

$$\left(N\overline{Y} | Y_1, \ldots, Y_n, X, \theta\right) \sim N\left(n\bar{y} + \beta \sum_{n+1}^{N} X_i, \sigma^2 \sum_{n+1}^{N} X_i^{2g}\right). \quad (2.5.13)$$

Next we find the posterior distribution of β given σ. Since

$$(Y_i | X, \theta) \sim N\left(\beta X_i, \sigma^2 X_i^{2g}\right),$$

it follows that

$$(Y_i X_i^{-g} | X, \theta) \sim N\left(\beta X_i^{1-g}, \sigma^2\right),$$

so that $Y_i X_i^{-g}$ has a simple regression (through the origin) on X_i^{1-g}. In this case with $\Pr(\beta|\sigma) \propto \text{const}$, the conditional distribution of β given σ and $(Y_1, \ldots, Y_n)$ is normal, and standard least-squares computations provide the mean and variance (e.g., Box and Tiao, 1973, p. 115):

$$(\beta | X, Y_1, \ldots, Y_n, \sigma) \sim N\left(\hat{\beta}, \sigma^2 / \sum_{1}^{n} X_i^{2-2g}\right), \quad (2.5.14)$$

where $\hat{\beta}$ is given by (2.5.11). Combining (2.5.12) with (2.5.13), it follows that given σ and $(Y_1, \ldots, Y_n)$, $N\overline{Y}$ is normally distributed:

$$\left(N\overline{Y} | Y_1, \ldots, Y_n, \sigma\right)$$

$$\sim N\left(n\bar{y} + \hat{\beta} \sum_{n+1}^{N} X_i, \sigma^2 \left[\sum_{n+1}^{N} X_i^{2g} + \left(\sum_{n+1}^{N} X_i\right)^2 / \sum_{1}^{n} X_i^{2-2g}\right]\right). \quad (2.5.15)$$

When the prior distribution on $\log \sigma$ is proportional to a constant, $\sigma^2/\hat{\sigma}^2$ given $(Y_1, \ldots, Y_n)$ has an inverted χ^2 distribution on $n - 1$ degrees of freedom, where $\hat{\sigma}^2$ is given by (2.5.12), the usual residual mean-squares estimate from the regression of $Y_i X_i^{-g}$ on X_i^{1-g}. Averaging (2.5.15) over the distribution of σ^2 given $(Y_1, \ldots, Y_n)$ completes the proof (see also Box and Tiao, 1973, p. 117, and Ericson, 1969, Section 5, in particular).

2.6. BAYESIAN INFERENCE FOR A POPULATION QUANTITY

Let the objective of a survey be the estimation of a function of Y, and possibly also of X, in the population, say $Q = Q(X, Y)$. For example, Q could be the vector of mean incomes of families in each municipality, where income is a column of Y and municipality is encoded in a column of X. The quantity Q can be, without loss of generality, an exchangeable (possibly vector) function of the unit indices because the indexing of units, being random, implies that any population characteristic of interest must take the same functional form for all permutations of the indices. With scalar Y_i, a common quantity of interest is the population mean $\overline{Y} = \sum_1^N Y_i/N$.

Notation

It helps to have a convenient notation to refer to the various components of Y and R that repeatedly appear in expressions. Let $inc = \{(i, j)|I_{ij} = 1\}$ so that Y_{inc} indicates the components of Y *inc*luded in the sample and R_{inc} indicates the components of R included in the sample; analogously define $exc = \{(i, j)|I_{ij} = 0\}$. In a slight abuse of standard notation, we let $Y = (Y_{inc}, Y_{exc})$ and $R = (R_{inc}, R_{exc})$. Both Y_{exc} and R_{exc} are unobserved, since they are *exc*luded from the sample; R_{inc} is always fully observed, but Y_{inc} is only fully observed if there is no nonresponse.

Let $obs = \{(i, j)|I_{ij} = 1 \text{ and } R_{ij} = 1\}$ so that Y_{obs} indicates the components of Y that are *obs*erved, that is, the components of Y_{inc} with associated $R_{ij} = 1$; analogously define $mis = \{(i, j)|I_{ij} = 1 \text{ and } R_{ij} = 0\}$ so that Y_{mis} indicates the components of Y that are *mis*sing, that is, the components of Y_{inc} with associated $R_{ij} = 0$, $Y_{inc} = (Y_{obs}, Y_{mis})$. Finally, let $nob = \{(i, j)|I_{ij} = 0 \text{ or } R_{ij} = 0\}$ so that $Y_{nob} = (Y_{exc}, Y_{mis})$ indicates the *n*ot *ob*served components of Y; $Y = (Y_{obs}, Y_{nob})$.

When Y_i and R_i are scalars, so that the j subscript is irrelevant, this notation refers in the obvious manner to the set of i subscripts; for example, $mis = \{i|I_i = 1 \text{ and } R_i = 0\}$.

The Posterior Distribution for $Q(X, Y)$

A Bayesian inference for $Q(X, Y)$ follows from its posterior distribution, that is, its conditional distribution given the observed values (X, Y_{obs}, R_{inc}, I) calculated under specified models. Using the notation we have established, we write this posterior distribution as $\Pr(Q|X, Y_{obs}, R_{inc}, I)$. For example, if the posterior distribution for scalar Q is normal with mean $\hat{Q} = \hat{Q}(X, Y_{obs}, R_{inc}, I)$ and variance $U =$

$U(X, Y_{obs}, R_{inc}, I)$, then an interval that includes Q with 95% probability is $\hat{Q} \pm 1.96 U^{1/2}$.

Specific examples of the conditional distribution of $\overline{Y}$ given X and $Y_{obs} = (Y_1, \ldots, Y_n)$ were given in Section 2.5. When $obs = \{1, \ldots, n\}$ and the extra conditioning on R_{inc} and I in $\Pr(Q|X, Y_{obs}, R_{inc}, I)$ can be ignored, these examples thus provide illustrations of Bayesian calculations for the posterior distribution of $Q = \overline{Y}$. We see in this section that this extra conditioning on R_{inc} and I can be ignored when both the sampling and response mechanisms are unconfounded.

Relating the Posterior Distribution of Q to the Posterior Distribution of Y_{nob}

Since $Q(X, Y)$ is a function of observed values in (X, Y_{obs}) and unobserved values in Y_{nob}, the posterior distribution of Q can be calculated from the observed values (X, Y_{obs}) and the posterior distribution of Y_{nob}:

$$\Pr(Q|X, Y_{obs}, R_{inc}, I)$$

$$= \int_{\mathcal{Y}(Q)} \Pr(Y_{nob}|X, Y_{obs}, R_{inc}, I) \, dY_{nob} \qquad (2.6.1)$$

where

$$\mathcal{Y}(Q) = \{Y_{nob}|Q(X, Y) = Q'\}.$$

General expressions for the posterior distribution of Y_{nob} are simpler than those for Q since they do not involve the integration called for in equation (2.6.1). Consequently, we will proceed by considering the posterior distribution of Y_{nob} rather than that of Q, keeping in mind that the posterior distribution of Q can then be obtained from (2.6.1), which in particular cases is often easy to evaluate.

The posterior distribution of Y_{nob} can be written directly in terms of the specifications for (X, Y) and the sampling and response mechanisms as

$$\Pr(Y_{nob}|X, Y_{obs}, R_{inc}, I)$$

$$= \frac{\int \Pr(X, Y) \Pr(R|X, Y) \Pr(I|X, Y, R) \, dR_{exc}}{\iint \Pr(X, Y) \Pr(R|X, Y) \Pr(I|X, Y, R) \, dR_{exc} \, dY_{nob}}, \qquad (2.6.2)$$

where the probabilities are evaluated at the observed values of X, Y_{obs}, R_{inc} and I; thus, in (2.6.2) and throughout this section, the value of (X, Y_{obs}, R_{inc}, I) is fixed, as are the values of the sets *inc*, *obs*, *nob*, and *mis*. Throughout the text, when one of these sets of indices appears explicitly as the subscript of a variable being conditioned upon, that set of indices is implicitly being fixed.

Ignorable Sampling Mechanisms

The sampling mechanism is defined to be ignorable at the observed values (X, Y_{obs}, R_{inc}, I) if

$$\Pr(Y_{nob}|X, Y_{obs}, R_{inc}, I) = \frac{\int \Pr(X, Y)\Pr(R|X, Y)\, dR_{exc}}{\iint \Pr(X, Y)\Pr(R|X, Y)\, dR_{exc}\, dY_{nob}}. \quad (2.6.3)$$

That is, the sampling mechanism is ignorable if the posterior distribution of Y_{nob} (and thus of Q) does not explicitly involve the particular specification for the sampling mechanism; in other words, for a fixed observed value of (X, Y_{obs}, R_{inc}, I), and fixed specifications for $\Pr(X, Y)$ and $\Pr(R|X, Y)$, the posterior distribution of Y_{nob} will be the same for all ignorable sampling mechanisms. Since the right-hand side of (2.6.3) is the conditional distribution of Y_{nob} given (X, Y_{obs}, R), (2.6.3) is the same as

$$\Pr(Y_{nob}|X, Y_{obs}, R_{inc}, I) = Pr(Y_{nob}|X, Y_{obs}, R_{inc}). \quad (2.6.4)$$

It is important to keep in mind that the sets *nob*, *obs*, *exc*, and *inc*, which are functions of I and appear in the right-hand sides of (2.6.3) and (2.6.4), are fixed at their observed values even though I is not explicitly conditioned upon in these expressions.

Result 2.1. An Equivalent Definition for Ignorable Sampling Mechanisms

The sampling mechanism is ignorable at the observed values (X, Y_{obs}, R_{inc}, I) if and only if

$$\Pr(I|X, Y, R_{inc}) = \Pr(I|X, Y_{obs}, R_{inc}) \quad \text{for all possible } Y_{nob}. \quad (2.6.5)$$

The equivalence of (2.6.5) and (2.6.4) follows immediately from Bayes's theorem applied to the joint distribution of I and Y_{nob} given (X, Y_{obs}, R_{inc}):

$$\frac{\Pr(I|X, Y, R_{inc})}{\Pr(I|X, Y_{obs}, R_{inc})} = \frac{\Pr(Y_{nob}|X, Y_{obs}, R_{inc}, I)}{\Pr(Y_{nob}|X, Y_{obs}, R_{inc})}.$$

BAYESIAN INFERENCE FOR A POPULATION QUANTITY

If the sampling mechanism is unconfounded, (2.4.1) holds for all values of (X, Y, R, I), and thus in particular (2.6.5) holds. Thus an unconfounded probability sampling mechanism is always ignorable. Confounded sampling mechanisms may also be ignorable, however. For example, a sequential sampling mechanism, where the decision about how many units to sample is based on an analysis of observed data from initial units in the sample, is confounded but is ignorable since $\Pr(I|X, Y, R)$ depends only on (X, Y_{obs}, R_{inc}).

Ignorable Response Mechanisms

The response mechanism is said to be ignorable at the observed values (X, Y_{obs}, R_{inc}, I) if

$$\Pr(Y_{nob}|X, Y_{obs}, R_{inc}, I) = \frac{\int \Pr(X, Y)\Pr(I|X, Y, R)\, dR_{exc}}{\iint \Pr(X, Y)\Pr(I|X, Y, R)\, dR_{exc}\, dY_{nob}}.$$

(2.6.6)

That is, the response mechanism is ignorable if the posterior distribution of Y_{nob} does not explicitly involve the particular specification for the response mechanism; in other words, for a fixed observed value of (X, Y_{obs}, R_{inc}, I), and fixed specifications for $\Pr(X, Y)$ and $\Pr(I|X, Y, R)$, the posterior distribution of Y_{nob} will be the same for all ignorable response mechanisms.

Result 2.2. Ignorability of the Response Mechanism When the Sampling Mechanism Is Ignorable

Suppose the sampling mechanism is ignorable at (X, Y_{obs}, R_{inc}, I). Then the response mechanism is ignorable at (X, Y_{obs}, R_{inc}, I) if and only if

$$\Pr(R_{inc}|X, Y) = \Pr(R_{inc}|X, Y_{obs}). \quad (2.6.7)$$

The proof of this result is straightforward. Since the sampling mechanism is ignorable, by (2.6.5) the right-hand side of (2.6.6) equals $\Pr(X, Y)/\int \Pr(X, Y)\, dY_{nob}$, which is $\Pr(Y_{nob}|X, Y_{obs})$. Thus, with ignorable sampling mechanisms, the definition of an ignorable response mechanism is

$$\Pr(Y_{nob}|X, Y_{obs}, R_{inc}, I) = \Pr(Y_{nob}|X, Y_{obs}), \quad (2.6.8)$$

or, by (2.6.4)

$$\Pr(Y_{nob}|X, Y_{obs}, R_{inc}) = \Pr(Y_{nob}|X, Y_{obs}),$$

which is equivalent to (2.6.7) by Bayes's theorem:

$$\frac{\Pr(Y_{nob}|X, Y_{obs}, R_{inc})}{\Pr(Y_{nob}|X, Y_{obs})} = \frac{\Pr(R_{inc}|X, Y)}{\Pr(R_{inc}|X, Y_{obs})}.$$

Result 2.3. The Practical Importance of Ignorable Mechanisms

When both the sampling and response mechanisms are ignorable, the posterior distribution of Y_{nob} (and thus of Y_{mis} and Q) can be obtained from the observed values and the specification for $\Pr(X, Y)$:

$$\Pr(Y_{nob}|X, Y_{obs}, R_{inc}, I) = \Pr(Y_{nob}|X, Y_{obs})$$

$$= \Pr(X, Y) \Big/ \int \Pr(X, Y) \, dY_{nob}. \quad (2.6.9)$$

The proof is immediate from (2.6.8).

Drawing inferences from (2.6.9) ignores the processes that create missing data, that is, the sampling and response mechanisms. By Result 2.1, an unconfounded sampling mechanism is ignorable, and confounded but well-designed sampling mechanisms are also ignorable. Consequently, it is generally appropriate to regard the sampling mechanism as ignorable in scientific surveys. If, in addition, the response mechanism is unconfounded, it too is ignorable; that is, since

$$\Pr(R_{inc}|X, Y) = \int \Pr(R|X, Y) \, dR_{exc}$$

an unconfounded response mechanism satisfies (2.6.7), and thus Result 2.2 implies that the response mechanism is ignorable. Furthermore, confounded response mechanisms can be ignorable. For example, when nonresponse on the first component of bivariate Y_i depends on the second component, which is always observed, nonresponse is ignorable by (2.6.7). But generally it will not be appropriate to assume automatically an ignorable response mechanism, and thus not appropriate to draw inferences using (2.6.9); rather, generally we must use (2.6.2), or (2.6.3) if the sampling mechanism is ignorable.

Relating Ignorable Sampling and Response Mechanisms to Standard Terminology in the Literature on Parametric Inference from Incomplete Data

Both the sampling and the response mechanisms can be considered to be processes that create missing data. Within the literature on parametric inference with missing data, the standard structure posits one model for the data given a parameter θ, say $f(X, Y|\theta)$, and another model for the missing-data indicators given (X, Y) and a missingness parameter ϕ, say $g(I, M|X, Y, \phi)$ where in terms of our notation, the missingness indicator $M = \{M_{ij}\} = \{I_{ij}R_{ij}\}$ is fully observed like I, with (I, M) and (I, R_{inc}) one–one functions of each other. When the function $g(I, M|X, Y, \phi)$ evaluated at the observed values of I, M, X and the observed components of Y does not depend on missing components of Y, Y_{nob}, then the missing data are said to be *missing at random* (Rubin, 1976a). When in addition, θ and ϕ are *distinct*—that is, *a priori* independent—then Theorem 5.2 in Rubin (1976a) shows that Bayesian inference can ignore the process that creates missing data and estimate θ from $f(X, Y|\theta)$, the prior distribution on θ, and the observed values (X, Y_{obs}).

Within the context of Bayesian inference for Y, the combination of the missing data being missing at random and θ and ϕ being distinct is practically equivalent to ignorable sampling and response mechanisms. To see this, we first show that if θ and ϕ are *a priori* independent and the missing data are missing at random, then the sampling and response mechanisms are ignorable. In general

$$\Pr(Y_{nob}|X, Y_{obs}, R_{inc}, I) = \Pr(Y_{nob}|X, Y_{obs}, M, I),$$

which, since θ and ϕ are *a priori* independent, can be written as

$$\Pr(Y_{nob}|X, Y_{obs}, R_{inc}, I)$$
$$= \frac{\iint f(X, Y|\theta) g(I, M|X, Y, \phi) \Pr(\theta) \Pr(\phi) \, d\theta \, d\phi}{\iiint f(X, Y|\theta) g(I, M|X, Y, \phi) \Pr(\theta) \Pr(\phi) \, d\theta \, d\phi \, dY_{nob}}.$$

But since the missing data are missing at random, $g(I, M|X, Y, \phi) = g(I, M|X, Y_{obs}, \phi)$, and we have that

$$\Pr(Y_{nob}|X, Y_{obs}, R_{inc}, I) = \frac{\int f(X, Y|\theta) \Pr(\theta) \, d\theta}{\iint f(X, Y|\theta) \Pr(\theta) \, d\theta \, dY_{nob}}$$

$$= \Pr(Y_{nob}|X, Y_{obs}),$$

which is the definition of ignorable sampling and response mechanisms. Next, note that if the sampling and response mechanisms are ignorable and a particular $\Pr(X, Y)$ is specified for drawing inferences for Y, we can choose $f(X, Y|\theta)$ and $\Pr(\theta)$ such that

$$\int f(X, Y|\theta)\Pr(\theta)\, d\theta = \Pr(X, Y)$$

and then choose $g(I, M|X, Y, \phi)$ such that the missing data are missing at random with ϕ distinct from θ.

2.7. INTERVAL ESTIMATION

An interval estimate of Q is a region of k-dimensional space, where k is the dimension of Q, that (a) is a function only of observed values, (X, Y_{obs}, R_{inc}, I) and (b) includes Q with a specified coverage rate between 0 and 1, where the definition of coverage depends on the mode of inference. In many cases, the interval is centered at a point estimate of Q, and the coverage is set high enough so that in senses to be described, it is highly likely that the interval includes Q. For example, the standard nominal 95% interval estimate of scalar $\overline{Y}$ in a simple random sample with no possibility of nonresponse is

$$\bar{y} \pm 1.96 s (n^{-1} - N^{-1})^{1/2} \qquad (2.7.1)$$

where $n = \sum_1^N I_i =$ the sample size, $\bar{y} = \sum_1^N I_i Y_i / n =$ the sample mean, and $s^2 = \sum_1^N I_i (Y_i - \bar{y})^2 / (n - 1) =$ the sample variance.

More generally, interval estimates of Q are often created from (a) a point estimate of Q, $\hat{Q} = \hat{Q}(X, Y_{obs}, R_{inc}, I)$, (b) a statistic measuring the variance of $Q - \hat{Q}$, $T = T(X, Y_{obs}, R_{inc}, I)$, and (c) an assumption of normality:

$$(Q - \hat{Q}) \sim N(0, T)$$

where for Q with k components, $\hat{Q}$ has k components and T is a $k \times k$ matrix. The nominal coverage of an interval estimate of Q centered at $\hat{Q}$ is then the integral of the $N(0, T)$ distribution over the region defined by the interval. With interval (2.7.1), k is 1, Q is $\overline{Y}$, $\hat{Q}$ is $\bar{y}$, T is $s^2(n^{-1} - N^{-1})$, and the nominal coverage is the integral from $-1.96T^{1/2}$ to $+1.96T^{1/2}$ of the $N(0, T)$ distribution—essentially 95%.

INTERVAL ESTIMATION

General Interval Estimates

In full generality, let $C = C(X, Y_{obs}, R_{inc}, I)$ be an interval estimate of Q with coverage $1 - \alpha$, $0 \leq \alpha \leq 1$, and let $\delta(Q \in C)$ be the indicator for whether C includes Q:

$$\delta(Q \in C) = \begin{cases} 1 & \text{if } Q \in C, \\ 0 & \text{otherwise.} \end{cases}$$

Then the coverage $1 - \alpha$ satisfies

$$E[\delta(Q \in C)|A] = 1 - \alpha \qquad (2.7.2)$$

where the values being conditioned upon, A, depend on the mode of inference; for Bayesian inference, A equals the observed values (X, Y_{obs}, R_{inc}, I), whereas for randomization-based inference, A equals the fixed population values (X, Y, R) or possibly just (X, Y). Thus the coverage of an interval estimate of Q is the conditional probability that the interval includes Q.

Bayesian Posterior Coverage

The Bayesian posterior coverage of C for Q is defined to be the posterior probability that $Q \in C$:

$$\text{Prob}\{Q \in C | X, Y_{obs}, R_{inc}, I\} = E[\delta(Q \in C)|X, Y_{obs}, R_{inc}, I], \quad (2.7.3)$$

or

$$\text{Prob}\{Q \in C | X, Y_{obs}, R_{inc}, I\} = \int \delta(Q \in C) \text{Pr}(Q|X, Y_{obs}, R_{inc}, I) \, dQ,$$

$$(2.7.4)$$

where $\text{Prob}\{B|A\}$ is the conditional probability of event B given event A. When the sampling mechanism is ignorable, $\text{Pr}(Q|X, Y_{obs}, R_{inc}, I)$ in (2.7.4) can be replaced by $\text{Pr}(Q|X, Y_{obs}, R_{inc})$, with the understanding that the sets *obs* and *inc* are being implicitly fixed even though I is not being explicitly conditioned upon. The posterior coverage probability of C can also be written directly in terms of the posterior distribution of Y_{nob} using (2.6.1):

$$\text{Prob}(Q \in C | X, Y_{obs}, R_{inc}, I)$$

$$= \int \delta(Q \in C) \int_{\mathcal{Y}(Q)} \Pr(Y_{nob} | X, Y_{obs}, R_{inc}, I) \, dY_{nob} \, dQ$$

$$= \int \delta(Q \in C) \Pr(Y_{nob} | X, Y_{obs}, R_{inc}, I) \, dY_{nob}.$$

Since $Q = Q(X, Y)$ depends on unobserved Y_{nob}, and $C = C(X, Y_{obs}, R_{inc}, I)$ involves only observed values, the random variable in the expectation in (2.7.3) used to define the posterior probability coverage is Q, the quantity to be estimated. Posterior probability coverage depends on the observed values, the particular interval C, the quantity to be estimated Q, and the joint specification for the random variables (X, Y, R, I).

A Bayesian $100(1 - \alpha)\%$ posterior interval estimate of Q has the direct frequency interpretation that in $100(1 - \alpha)\%$ of the situations having (i) the same essential survey conditions (as summarized by the model specifications) and (ii) the same observed values, Q will be included in the interval.

Example 2.5. Interval Estimation in the Context of Example 2.1

Consider Example 2.1 with $Y_{obs} = (Y_1, \ldots, Y_n)$, and suppose ignorable sampling and response mechanisms that yield $n > 1$, so that the posterior distribution of $\bar{Y}$ is t on $n - 1$ degrees of freedom with location $\bar{y}$ and scale $s(n^{-1} - N^{-1})^{1/2}$. Then the interval

$$\bar{y} \pm t_{n-1}(\alpha/2) s(n^{-1} - N^{-1})^{1/2}, \tag{2.7.5}$$

where $t_{n-1}(\alpha/2)$ is the upper $100\alpha/2$ percentage point of the standard t distribution on $n - 1$ degrees of freedom, has constant probability coverage $1 - \alpha$ for all possible observations. The t distribution for samples with n modestly large, say greater than 10, can be approximated fairly well by a normal distribution so that the probability coverage of interval (2.7.1) is close to 95% in such cases.

Fixed-Response Randomization-Based Coverage

Posterior probability coverage is not the measure of coverage that most survey practitioners have been trained to use. Commonly, survey practi-

INTERVAL ESTIMATION

tioners have been trained using randomization-based inference, as described in Cochran (1977), Hansen, Hurwitz, and Madow (1953), Kish (1965), and other more recent textbooks on the practice of sample surveys. In this approach, coverage is defined in terms of performance in repeated drawings from the same population using the same unconfounded probability sampling mechanism. Generally, the fixed-response randomization-based coverage of interval $C = C(X, Y_{obs}, R_{inc}, I)$ for Q is defined to be

$$\text{Prob}\{Q \in C | X, Y, R\} = E[\delta(Q \in C) | X, Y, R]$$

$$= \int \delta(Q \in C) \text{Pr}(I | X, Y, R) \, dI, \quad (2.7.6)$$

where $\text{Pr}(I|X)$ rather than $\text{Pr}(I|X, Y, R)$ can be used in (2.7.6) when the sampling mechanism is unconfounded; $\text{Prob}\{Q \in C | X, Y, R\}$ in (2.7.6) is called the fixed-response randomization-based coverage of C for Q because (a) R, response, is being treated as fixed, just as is (X, Y), and (b) the coverage is over the possible randomizations defined by the sampling mechanism. Notice from (2.7.6) that in contrast to posterior probability coverage, with randomization-based coverage the random variable in the expectation is the interval C—which depends on I, rather than the quantity Q—which is being treated as fixed since it involves only (X, Y). Fixed-response randomization-based coverage depends on Q, C, the specification for the sampling mechanism, and the specific values of (X, Y, R).

Fixed-response randomization-based coverage directly addresses the following question: suppose the population is such that (X, Y, R) takes on specific hypothesized values, and we intend to estimate Q using interval C from samples drawn using sampling mechanism $\text{Pr}(I|X, Y, R)$—then what is the probability that C will include Q? If for the chosen $\text{Pr}(I|X, Y, R)$, $\text{Prob}\{Q \in C | X, Y, R\}$ is large, say at least 95%, for a range of hypothesized values of (X, Y, R) considered reasonable at the design stage, then the investigator who decided to estimate Q using C from the survey would be at least 95% confident that the value of C to be observed would include Q. In such a case, the observed value of C is called a 95% confidence interval for Q.

With no possibility of nonresponse ($R_{ij} \equiv 1$), interval (2.7.1) is generally considered a 95% confidence interval for scalar $\overline{Y}$ when the population Y_i values are not long-tailed and simple random sampling is being used. That is, it is commonly asserted that in 95% of the $\binom{N}{n}$ possible simple random samples from a fixed population, the observed intervals $\bar{y} \pm 1.96s(n^{-1} - N^{-1})^{1/2}$ will include the fixed but unknown $\overline{Y}$. The justification for interval estimates of this form is an appeal to the central limit theorem effect with large samples from large populations (Madow, 1948; Hájek, 1960).

Random-Response Randomization-Based Coverage

When nonresponse is possible, it is convenient to extend randomization inference to include a specification for the response mechanism, $\Pr(R|X, Y)$, and then define the coverage of C for Q as

$$\text{Prob}\{Q \in C|X, Y\} = E[\delta(Q \in C)|X, Y]$$

$$= \iint \delta(Q \in C)\Pr(I|X, Y, R)\Pr(R|X, Y)\, dI\, dR$$

$$= \int \text{Prob}\{Q \in C|X, Y, R\}\Pr(R|X, Y)\, dR \quad (2.7.7)$$

where $\Pr(I|X, Y, R)$ can be replaced by $\Pr(I|X)$ if the sampling mechanism is unconfounded, and $\Pr(R|X, Y)$ can be replaced by $\Pr(R|X)$ if the response mechanism is unconfounded; $\text{Prob}\{Q \in C|X, Y\}$ in (2.7.7) is called the random-response randomization-based coverage of C for Q. Again, C is the random variable in the expectation of $\delta(Q \in C)$; $Q = Q(X, Y)$ is fixed since X and Y are fixed.

Random-response randomization-based coverage directly addresses the following question: suppose the population is such that (X, Y) takes on specific hypothesized values and nonresponse is created according to the response mechanism $\Pr(R|X, Y)$; further suppose that we intend to estimate Q using interval C from samples drawn using the sampling mechanism $\Pr(I|X, Y, R)$—what is the probability that C will include Q? If for the chosen $\Pr(I|X, Y, R)$ and the hypothesized $\Pr(R|X, Y)$, $\text{Prob}\{Q \in C|X, Y\}$ is at least 95% for a range of values of (X, Y) considered reasonable at the design stage, then assuming the propriety of the specification for the response mechanism, the investigator who decides to estimate Q using C from the survey would be at least 95% confident that the value of C to be observed would include Q. In such a case, the observed value of C is called a 95% random-response confidence interval for Q.

Nominal versus Actual Coverage of Intervals

In applied inference, interval estimates nearly always should be regarded as approximate in the sense that their associated coverages are nominal and not exactly correct. For instance, we have seen that interval (2.7.1) can be regarded as a nominal 95% Bayesian probability interval for $\overline{Y}$ under the conditions of Example 2.5 by approximating the t posterior distribution of Q by a normal distribution. In fact, interval (2.7.1) is often regarded as a nominal 95% Bayesian probability interval for $\overline{Y}$ when no particular specification for $\Pr(Y)$ is made; the justification for such an assertion is the result in Example 2.2, concerning posterior moments, combined with the discussion following Example 2.3 and the normal approximation to the posterior distribution motivated in Section 2.10. In the absence of nonresponse, we have seen that the same interval (2.7.1) is commonly regarded as a nominal 95% randomization-based interval for Y in simple random samples, even though the exact randomization-based coverage of the interval will depend on the exact values in Y. In other cases, interval estimates for Q may be derived ignoring the response mechanism even when it is known that this assumption is not entirely appropriate; the extra effort required to build an appropriate nonignorable response mechanism and analyze data under it might not be considered to be worth the effort.

2.8. BAYESIAN PROCEDURES FOR CONSTRUCTING INTERVAL ESTIMATES, INCLUDING SIGNIFICANCE LEVELS AND POINT ESTIMATES

There are two rather standard Bayesian procedures for constructing interval estimates of a population quantity Q. The first fixes the coverage rate $1 - \alpha$ in advance and determines C to include the most likely values of Q totalling $1 - \alpha$ posterior probability, where "most likely" is usually defined as highest posterior density. The second procedure fixes a null value of Q in advance, say Q_0, determines C by the collection of values of Q more likely than Q_0, and calculates the coverage $1 - \alpha$ as the posterior probability of this C; the resultant α is commonly called the significance level of the null value Q_0.

Highest Posterior Density Regions

Very commonly interval estimates of Q are designed to include the most likely values of Q, values within the interval being considered plausible and values outside the interval being considered implausible. The standard 95% interval for $\overline{Y}$, expression (2.7.1), is an example of such an interval. The standard Bayesian procedure for creating an interval of this type is to fix $1 - \alpha$ in advance, say at 95%, and determine C as the collection of values of Q such that (i) the posterior probability that $Q \in C$ is $1 - \alpha$, and

(ii) every point in C has higher posterior density than every point outside C:

$$\int \delta(Q \in C) \Pr(Q|X, Y_{obs}, R_{inc}, I) \, dQ = 1 - \alpha,$$

and (2.8.1)

$$\Pr(Q'|X, Y_{obs}, R_{inc}, I) > \Pr(Q''|X, Y_{obs}, R_{inc}, I)$$

for every $Q' \in C$ and $Q'' \notin C$. Or using a simplified notation

$$\text{Prob}\{Q \in C | X, Y_{obs}, R_{inc}, I\} = 1 - \alpha$$

and (2.8.2)

$$d(Q') > d(Q'') \quad \text{for every} \quad Q' \in C \text{ and } Q'' \notin C$$

where $d(Q)$ is the posterior density at Q, $d(Q) = \Pr(Q|X, Y_{obs}, R_{inc}, I)$.

When the posterior distribution of the k-dimensional row vector Q is normal with mean $\hat{Q}$ and variance T, highest posterior density regions are easily constructed as follows: for fixed coverage $1 - \alpha$, C is the set of all Q such that

$$(Q - \hat{Q}) T^{-1} (Q - \hat{Q})^t < \chi_k^2(\alpha) \qquad (2.8.3)$$

where $\chi_k^2(\alpha)$ is the 100α upper percentage point of the chi-squared distribution on k degrees of freedom. This follows from basic distribution theory because if $(Q - \hat{Q}) \sim N(0, T)$ where Q is k-dimensional, then $(Q - \hat{Q}) T^{-1} (Q - \hat{Q})^t$ is distributed as χ^2 on k degrees of freedom.

More generally, suppose $(Q - \hat{Q})$ is multivariate t with scale $T^{1/2}$ and ν degrees of freedom. Then, for fixed coverage $1 - \alpha$, C is the set of all Q such that

$$(Q - \hat{Q}) T^{-1} (Q - \hat{Q})^t < k F_{k,\nu}(\alpha) \qquad (2.8.4)$$

where $F_{k,\nu}(\alpha)$ is the 100α upper percentage point of the F distribution with k and ν degrees of freedom.

More discussion of such highest posterior density regions is given in Box and Tiao (1973, pp. 121–126).

Significance Levels—p-Values

In many statistical analyses there is one value of the quantity $Q(X, Y)$ that is of particular interest. For example, let Q be the vector of male–female differences in average incomes across 10 occupational strata in the population. Suppose it is of interest to summarize evidence concerning the null value $Q_0 = (0, \ldots, 0)$ representing no difference in average incomes within strata. From the Bayesian perspective, one way to summarize such evidence is to calculate a highest posterior density region for Q with α fixed in advance and C determined by the most likely values of Q, and see whether or not Q_0 is inside this region. A closely related method creates an interval estimate of Q where neither the region C nor the coverage α is fixed in advance, but both are determined by Q_0 and the posterior distribution of Q. Specifically, partition the possible values of Q into (i) those values of Q that have higher posterior density than Q_0 and (ii) those values of Q that have posterior density no higher than Q_0, and then calculate the posterior probability that Q is in the first group. Using the simplified notation in (2.8.2), for fixed Q_0, C and then α are determined by

$$C = \{Q | d(Q) > d(Q_0)\} \quad \text{and} \quad \text{Prob}\{C | X, Y_{obs}, R_{inc}, I\} = 1 - \alpha$$

or leaving the definition of C implicit,

$$\text{Prob}\{d(Q) > d(Q_0) | X, Y_{obs}, R_{inc}, I\} = 1 - \alpha. \tag{2.8.5}$$

The higher this posterior probability, $1 - \alpha$, the less believable is Q_0 or values near it; α can be called the Bayesian significance level or *p*-value of the null value Q_0. Figure 2.2 depicts this quantity,

$$p\text{-value}\{Q_0 | X, Y_{obs}, R_{inc}, I\} = \text{Prob}\{d(Q) \leq d(Q_0) | X, Y_{obs}, R_{inc}, I\}.$$
$$\tag{2.8.6}$$

With normal posterior distributions, the significance level of Q_0 is given by

$$p\text{-value}\{Q_0 | X, Y_{obs}, R_{inc}, I\} = \text{Prob}\{\chi_k^2 > (Q_0 - \hat{Q}) T^{-1} (Q_0 - \hat{Q})'\}$$
$$\tag{2.8.7}$$

where χ_k^2 is a χ^2 random variable on k degrees of freedom. With t posterior distributions, the significance level of Q_0 is given by

$$p\text{-value}\{Q_0 | X, Y_{obs}, R_{inc}, I\}$$
$$= \text{Prob}\{F_{k,\nu} > (Q_0 - \hat{Q}) T^{-1} (Q_0 - \hat{Q})'/k\} \tag{2.8.8}$$

where $F_{k,\nu}$ is an F random variable on k and ν degrees of freedom.

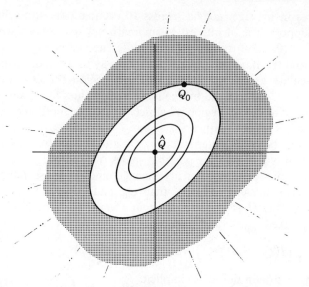

Figure 2.2. Contours of the posterior distribution of Q with the null value Q_0 indicated. The significance level of Q_0 is the posterior probability that Q is in the shaded area and beyond.

Point Estimates

A common applied objective is to produce one value for the unknown Q that "best" summarizes its unknown value. With symmetric unimodal posterior distributions, such as the normal or t, the obvious point estimate is the center of symmetry, which is the center of all $1 - \alpha$ highest posterior density intervals as well as the limit of those intervals as $\alpha \to 1$, that is, the posterior mode. With asymmetric posterior distributions, there is no such obvious choice, the posterior mean and median being common alternatives to the mode. To decide which point estimate is "best" requires a definition of "best," which usually entails consideration of a loss function (e.g., the posterior mean minimizes squared-error loss).

2.9. EVALUATING THE PERFORMANCE OF PROCEDURES

Suppose that C is an interval estimate of Q with nominal Bayes posterior probability coverage $1 - \alpha_b$, with nominal fixed-response randomization-based coverage $1 - \alpha_f$, and with nominal random-response randomization-based coverage $1 - \alpha_r$. How should these nominal levels be evaluated? With one sample from one population, the answers are: compare $1 - \alpha_b$ with $\text{Prob}\{Q \in C | X, Y_{obs}, R_{inc}, I\}$ under the correct specifications for

Pr(X, Y), Pr($R|X, Y$) and Pr($I|X, Y, R$); compare $1 - \alpha_f$ with Prob$\{Q \in C|X, Y, R\}$ under the correct specification for Pr($I|X, Y, R$) and the true values of X, Y, R in the population; and compare $1 - \alpha_r$ with Prob$\{Q \in C|X, Y\}$ under the correct specifications for Pr($R|X, Y$) and Pr($I|X, Y, R$) and the true values of X, Y in the population.

A Protocol for Evaluating Procedures

These comparisons, however, are not ideally suited for producing general summary advice concerning the propriety of the procedure used to construct the interval estimate since each comparison conditions on features specific to the particular sample or population. For a general evaluation of procedures for constructing intervals, a commonly accepted protocol is to (i) generate a sequence of populations and samples representative of those to which the procedure used to construct interval C might be applied to estimate Q, (ii) calculate the coverages Prob$\{Q \in C|X, Y_{obs}, R_{inc}, I\}$, Prob$\{Q \in C|X, Y, R\}$, and Prob$\{Q \in C|X, Y\}$ over the sequence, and (iii) see how close each coverage typically is to $1 - \alpha_b$, $1 - \alpha_f$, and $1 - \alpha_r$, respectively. The sequence of values of (X, Y) in the populations will be generated according to the specification for Pr(X, Y); then for each (X, Y) generated, nonresponse R will be generated according the response mechanism Pr($R|X, Y$); and finally, for each (X, Y, R), samples will be drawn according to the sampling mechanism Pr($I|X, Y, R$). Then Prob$\{Q \in C|X, Y_{obs}, R_{inc}, I\}$ will be calculated for each sample using the specifications used to generate the data; Prob$\{Q \in C|X, Y, R\}$ will be calculated for each population using the values of (X, Y, R) drawn for that population and the specification for Pr($I|X, Y, R$); and Prob$\{Q \in C|X, Y\}$ will be calculated for each population using the values of (X, Y) drawn for that population and the specifications for Pr($I|X, Y, R$) and Pr($R|X, Y$).

Ideally, the Bayesian would find that Prob$\{Q \in C|X, Y_{obs}, R_{inc}, I\} = 1 - \alpha_b$ for all samples; minimally, the Bayesian would hope that the average value of Prob($Q \in C|X, Y_{obs}, R_{inc}, I\}$ would be close to the nominal coverage of C, $1 - \alpha_b$. Similarly, ideally the fixed-response randomization-based coverage of C, Prob$\{Q \in C|X, Y, R\}$ would be identical to $1 - \alpha_f$ for all populations of (X, Y, R), but minimally, the average value of Prob$\{Q \in C|X, Y, R\}$ should be close to $1 - \alpha_f$. Finally, ideally, the random-response randomization-based coverage of C, Prob$\{Q \in C|X, Y\}$ should be close to $1 - \alpha_r$. If the nominal level of C equals the average coverage of C, the interval estimate is called *globally calibrated*, where the definition of coverage of C and thus calibration depends on the mode of inference. For example, the Bayesian's interval estimate is globally calibrated if $1 - \alpha_b = E[\text{Prob}\{Q \in C|X, Y_{obs}, R_{inc}, I\}]$. The following result implies that if a Bayesian, a fixed-response randomization-based advocate and a random-response randomization-based advocate all estimate Q using interval C with nominal coverages $1 - \alpha_b$, $1 - \alpha_f$, and $1 - \alpha_r$, respectively, and all are globally calibrated over the same sequence of events, then the nominal coverages are the same: $1 - \alpha_b = 1 - \alpha_f = 1 - \alpha_r = 1 - \alpha$.

Result 2.4. The Average Coverages Are All Equal to the Probability that C Includes Q

$$E[\text{Prob}\{Q \in C | X, Y_{obs}, R_{inc}, I\}] = E[\text{Prob}\{Q \in C | X, Y, R\}]$$
$$= E[\text{Prob}\{Q \in C | X, Y\}]$$
$$= \text{Prob}\{Q \in C\} \qquad (2.9.1)$$

The fact that the average coverages, no matter how coverage is defined, are equal is easy to prove. Simply recall that each coverage can be written as $E[\delta(Q \in C)|A]$ where $A = (X, Y_{obs}, R_{inc}, I)$, (X, Y, R), and (X, Y) for $\text{Prob}\{Q \in C | X, Y_{obs}, R_{inc}, I\}$, $\text{Prob}\{Q \in C | X, Y, R\}$ and $\text{Prob}\{Q \in C | X, Y\}$, respectively. Then realize that the average value of $E[\delta(Q \in C)|A]$ over repeated drawings from $\text{Pr}(I|X, Y, R)\text{Pr}(R|X, Y)\text{Pr}(X, Y)$ is simply $E\{E[\delta(Q \in C)]|A\} = E[\delta(Q \in C)] = \text{Prob}\{Q \in C\}$.

Further Comments on Calibration

As a consequence of Result 2.4, we can say that an interval estimate of Q, C with nominal coverage $1 - \alpha$ is globally calibrated if

$$1 - \alpha = \text{Prob}\{Q \in C\}, \qquad (2.9.2)$$

and not worry about which definition of coverage is being used. Similarly, we can say that the interval estimate is conservatively (globally) calibrated if $\text{Prob}\{Q \in C\} > 1 - \alpha$.

Such global calibration, averaging over the full sequence of possible populations and samples generated by $\text{Pr}(I|X Y, R)\text{Pr}(R|X, Y)\text{Pr}(X, Y)$, is actually a rather weak requirement. Ideally, an interval estimate would be calibrated for all identifiable sequences, that is, for all populations and samples that can be identified on the basis of observed values (X, Y_{obs}, R_{inc}, I). To see why, suppose C with nominal level 95% is globally calibrated, but in a particular situation that can be identified with Y_{obs}, is uncalibrated with coverage far less than 95%. When such a Y_{obs} occurs, it would be practically misleading to assert 95% coverage for C. For example, consider the nominal 95% interval (2.7.1) and suppose that Y_i is 0 or 1; when $\bar{y} = 0$, then interval (2.7.1) is simply the point 0, and it would usually be practically misleading to assert that the observed value of interval (2.7.1), $(0,0)$, has 95% coverage for $\bar{Y}$ even if interval (2.7.1) is nearly calibrated globally under $\Pr(I|X, Y, R)\Pr(R|X, Y)\Pr(X, Y)$.

Obviously, it would be ideal to be calibrated in all identifiable subsequences of samples and populations; that is, for the interval estimate C with nominal coverage $1 - \alpha$, ideally we could hope that

$$\text{Prob}\{Q \in C | X, Y_{obs}, R_{inc}, I\} = 1 - \alpha \qquad (2.9.3)$$

for all possible values of (X, Y_{obs}, R_{inc}, I). Since the left-hand side of (2.9.3) is the Bayesian posterior probability coverage of C, the Bayesian who does an exact analysis under the correct model to create C as a $100(1 - \alpha)\%$ interval for Q will be absolutely calibrated in the sense of (2.9.3). Without absolute calibration, there will, in principle, exist identifiable situations in which the nominal coverage will be too high or too low under the full specification $\Pr(X, Y)\Pr(R|X, Y)\Pr(I|X, Y, R)$. Nevertheless, because of the approximate nature of applied inference, absolute calibration is hopeless except possibly in the simplest situations. More realistically, we can hope to be calibrated conditional on some many–one function of the observed values, $h(X, Y_{obs}, R_{inc}, I)$, which summarizes important characteristics of the data:

$$\text{Prob}\{Q \in C | h(X, Y_{obs}, R_{inc}, I)\} = 1 - \alpha. \qquad (2.9.4)$$

The coarsest form of conditional calibration is simply global calibration (2.9.1) and the finest form is absolute calibration (2.9.3).

2.10. SIMILARITY OF BAYESIAN AND RANDOMIZATION-BASED INFERENCES IN MANY PRACTICAL CASES

The results and discussion of Section 2.9 suggest the idea that in many cases Bayesian and randomization-based inferences should be practically similar. For example, in a sample with no possibility of nonresponse and without covariates, the standard interval for $\bar{Y}$, $\bar{y} \pm 1.96s(n^{-1} - N^{-1})^{1/2}$, can be easily justified from the randomization perspective as an approximate 95% confidence interval and from the Bayesian perspective as an approximate 95% probability interval, as was seen in Examples 2.1–2.3 and suggested by Problems 7 and 10. The fact that many intervals in finite population inference besides $\bar{y} \pm 1.96s(n^{-1} - N^{-1})^{1/2}$ can be interpreted as either Bayesian probability intervals or randomization confidence intervals with the same nominal percent coverage, has been demonstrated in contexts with no possibility of nonresponse by Ericson (1969), Scott (1977), and Binder (1982), and illustrated here by Example 2.4 and Problems 11–14.

Standard Asymptotic Results Concerning Bayesian Procedures

Standard asymptotic results in mathematical statistics (see, e.g., Cox and Hinkley, 1974, Chapters 9 and 10 and their references) can be used to support the idea that interval estimates derived from the Bayesian perspective should be approximately calibrated from the random-response randomization perspective. The basic case for these standard results has $Z_1, \ldots, Z_n$ given θ as i.i.d. $h(\cdot|\theta)$ where θ_0 is the true value of θ. Given observed $Z_1, \ldots, Z_n$, let $\hat{\theta}$ and T be the posterior mean and variance (covariance) of θ, based on the correct model $h(\cdot|\theta)$ with a prior distribution on θ that has positive density around $\theta = \theta_0$.

Then under very weak regularity conditions, and in senses that can be made precise, as $n \to \infty$

$$\left((\theta - \hat{\theta})T^{-1/2}|Z_1, \ldots, Z_n\right) \to N(0, I) \qquad (2.10.1)$$

where I in (2.10.1) is the $k \times k$ identity matrix for θ with k components; thus the posterior distribution itself tends to be normal, with $(\theta - \hat{\theta})T^{-1/2}$ having a standard normal distribution given data $Z_1, \ldots, Z_n$.

Furthermore, over repeated samples of size n, (i) the distribution of the posterior mean tends to a normal distribution with mean θ_0;

$$\left(\hat{\theta}|\theta = \theta_0\right) \to N\left(\theta_0, V(\hat{\theta}|\theta = \theta_0)\right), \qquad (2.10.2)$$

and (ii) the distribution of the posterior variance tends to be centered at the variance of the posterior mean with lower-order variability than the posterior mean

$$(T|\theta = \theta_0) \to \left(V(\hat{\theta}|\theta = \theta_0), \ll V(\hat{\theta}|\theta = \theta_0)\right), \qquad (2.10.3)$$

where $A \to (B, \ll C)$ means that the distribution of A tends to be centered at B with each component having variability substantially less than each positive component of C.

As a consequence of (2.10.2) and (2.10.3), the sampling distribution of $(\hat{\theta} - \theta_0)T^{-1/2}$ under $h(\cdot|\theta = \theta_0)$ tends to a standard normal:

$$\left((\hat{\theta} - \theta_0)T^{-1/2}|\theta = \theta_0\right) \to N(0, I). \qquad (2.10.4)$$

Therefore, a Bayesian using (2.10.1) to assess the probability content of an interval estimate will arrive at the same values as the frequentist using (2.10.4) to evaluate the confidence coefficient of the interval estimate.

Extensions of These Standard Results

Actually, these conclusions hold under far more general conditions than the simple setup with i.i.d. observations from a model whose form is known.

SIMILARITY OF BAYESIAN AND RANDOMIZATION-BASED INFERENCES

Thus the normality of the posterior distribution in (2.10.1), and of the sampling distribution of the posterior mean in (2.10.4), hold in a very broad variety of other cases with the wrong model used for analysis. Also, the limiting equivalence of the posterior variance of θ and the sampling variance of the posterior mean in (2.10.3) holds very widely, and the expectation of the posterior mean under the true model is often asymptotically equal to the estimand of interest as in (2.10.2) (see, e.g., Huber, 1967, and interpret maximum-likelihood estimates as posterior means with normal posterior distributions).

The trick in applying these asymptotic results on i.i.d. sampling from an infinite population to our problem with a sampling mechanism, a response mechanism, and finite N is to imbed our variables $Z = (X, Y, R, I)$ in a sequence with $N \to \infty$, and formulate the sampling distributions of statistics using the random-response framework; a recent reference indicating the type of theory involved is Mitra and Pathak (1984).

Practical Conclusions of Asymptotic Results

The practical conclusion of the asymptotic results just presented is that both randomization and Bayesian inferences are usually based on the normal approximation $(Q - \hat{Q}) \sim N(0, T)$ with the same or similar statistics $\hat{Q}$ and T. Although the theoretical interpretations are different since the Bayesian statement is conditional on (X, Y_{obs}, R_{inc}, I) whereas the randomization-based statement is conditional on (X, Y), the practical inferences will thus often be essentially the same.

In a practical sense, a randomization inference that does not have an interpretation as an approximate Bayesian inference is not to be trusted inferentially since it does not then represent a fair assessment of the state of knowledge about the unknown Q under any model. Similarly, with standard scientific surveys, a Bayesian inference that does not have an interpretation as an approximate randomization inference is usually not to be trusted because one then could not describe an objective experiment with the population verifying that Bayesian inferences constructed in an analogous way are approximately calibrated.

Relevance to the Multiple-Imputation Approach to Nonresponse

The importance of these conclusions for the multiple-imputation approach to nonresponse is that they allow us usually to interpret an interval estimate for Q derived for a standard survey setting without nonresponse as both an approximate randomization confidence interval and an approximate Bayesian probability interval with the same nominal coverage. Most familiar complete-data inferences are justified by randomization theory arguments, but making the Bayesian interpretation allows us to use general and powerful Bayesian machinery to adjust the complete-data interval in the presence of nonresponse. More precisely, since the Bayesian position requires the specification of models for Pr(X, Y), the inclusion of a specific model for the response mechanism, Pr($R|X, Y$), creates no conceptual problem for the Bayesian as was seen in Section 2.6. In particular, given specified models, the Bayesian framework tells us how to create multiple imputations for nonresponse and how to average over the imputed values to create one final inference for Q. Again appealing to the dual interpretation of interval estimates, this final inference can then be viewed as either a Bayesian inference or a random-response randomization inference. This plan is theoretically justified in Chapters 3 and 4. Of course, this general perspective is only a guide for practice. Particular procedures created by following this pathway must be carefully evaluated in realistic situations before they can be unequivocally endorsed for practice.

PROBLEMS

1. Describe the difference between outcome variables Y and covariates X; give a specific example of a survey with both, and describe the purpose of recording the covariates.
2. Create examples of sampling and response mechanisms, both confounded and unconfounded. Provide descriptions of actual cases in which the mechanism might apply as well as mathematical statements of the mechanisms.
3. Give an example in which the assumption of stable response might not hold.
4. Prove equation (2.3.16).
5. Suppose $Y = (Y_1, \ldots, Y_N)^T$ is N-variate normal with mean 0, variance 1, and common correlation 0.1. Is this a unit exchangeable distribution and why? Introduce a parameter θ such that given θ the Y_i are i.i.d. What is the prior distribution of θ? Hint: Let $\theta \sim N(0, .1)$. Is this an example of de Finetti's theorem? Is there any difference if Y is N-variate normal with mean 0, variance 1, and common negative correlation? Discuss the loss in generality in assuming an i.i.d. model.
6. Obtain results parallel to those in Example 2.1 but for the case with $(\mu|\sigma^2) \sim N(\mu_0, \tau_0^2)$, μ_0, τ_0^2 fixed *a priori*. Comment on the sensitivity of the conditional distribution of $\bar{Y}$ given $(Y_1, \ldots, Y_n)$ to the values of μ_0, τ_0^2. Extend these results to the case with a proper inverted χ^2 prior distribution on σ^2.

PROBLEMS

7. Argue that the distribution of $\overline{Y}$ given $Y_1, \ldots, Y_n$ in Example 2.2 may be approximately normal. Consider the distribution of $\sum_{n+1}^{N} Y_i$ given θ and $(Y_1, \ldots, Y_n)$, then the averaging over the distribution of μ given σ^2 and $(Y_1, \ldots, Y_n)$, and finally the averaging over the distribution of σ^2 given $(Y_1, \ldots, Y_n)$.

8. Consider extending Example 2.2 to
   (a) higher moments of the posterior distribution of $\overline{Y}$,
   (b) percentiles of the posterior distribution of $\overline{Y}$, and
   (c) other functions of Y such as $\sum_1^N \log Y_i / N$ or median $\{Y_i\}$.

9. Extend Example 2.3 to include a proper Dirichlet prior distribution for the probabilities of each possible value of Y_i.

10. Summarize Pratt's (1965) arguments for the approximations
    $$E(\mu|Y_1, \ldots, Y_n) \doteq \bar{y},$$
    $$V(\mu|Y_1, \ldots, Y_n) \doteq s^2/n, \quad \text{and} \quad E(\sigma^2|Y_1, \ldots, Y_n) \doteq s^2.$$
    Can you find additional rationales for these approximations?

11. Show that $g = 1/2$ in Example 2.4 leads to the classical ratio estimator (Cochran, 1977, p. 151) and nearly to its associated estimate of scale [equation (6.9) in Cochran, 1977]. Describe the differences in the estimates of scale in the case of a large randomly drawn sample (so that $\overline{X}/\bar{x} \to 1$).

12. Show that $g = 1$ in Example 2.4 nearly leads (under appropriate conditions such as $N\bar{x} = 1$, $n/N \to 0$, $n \to \infty$, p.p.s. sampling) to the p.p.s. (probability proportional to size) estimator (Cochran, 1977, p. 253) and its associated estimate of scale [equation (9A.16) in Cochran, 1977]. Describe differences and situations where the resulting intervals for $\overline{Y}$ are nearly the same.

13. Show that $g = 0$ in Example 2.4 leads essentially to the regression through the origin estimator and its associated estimate of scale under simple random sampling.

14. Restrict Example 2.4 to $g = 0$ but extend it to consider q-variate X.
    (a) First, let $q = 2$ where $X_{i1} = 1$ for all i.
    (b) Second, let q be general.

15. Define a scalar estimand Q that is a function of all components of a multivariate Y_i and state why it might be of interest to estimate the quantity.

16. Describe the following sets: $inc \cup exc$, $mis \cup obs$, $exc \cap mis$, $mis \cap nob$, where $\cup$ is set union and $\cap$ is set intersection.

17. Relate ignorable sampling mechanisms and unconfounded sampling mechanisms. Describe a realistic sampling mechanism that is ignorable yet confounded.
18. Relate ignorable response mechanisms and unconfounded response mechanisms. Describe a realistic response mechanism with bivariate Y_i that is ignorable yet confounded (hint: let $R_{i2} \equiv 1$). Suppose that the sampling mechanism is unconfounded with R; show that if (2.6.7) holds, the response mechanism is ignorable and that $\Pr(Y_{nob}|X, Y_{obs}, R_{inc}, I) = \Pr(Y_{nob}|X, Y_{obs}, I)$.
19. Define a symmetric posterior interval for $\overline{Y}$ in the case of Example 2.4.
20. Consider the interval $(0, \infty)$. What is the posterior coverage of C in Example 2.1?
21. Should interval estimates for Q always be symmetric? Why or why not? Describe realistic situations in which symmetric intervals are appropriate and ones in which they are inappropriate.
22. Describe the essential difference between Bayesian posterior coverage and randomization-based coverage, both fixed-response and random-response. Discuss practical advantages and disadvantages of each definition of coverage. Similarly, describe the essential difference between fixed-response and random-response randomization-based coverage and practical advantages and disadvantages of each.
23. Can you think of other modes of randomization-based inference besides fixed-response and random-response?
    (a) For one example, might it be desirable to condition on some functions of I and R? Summarize Holt and Smith's (1979) arguments concerning conditional randomization inference. Also, use the empirical Bayes literature to suggest a family of estimators between the poststratified and simple random sample estimators.
    (b) For a second example, describe the Platek and Grey (1983) framework for addressing the effects of nonresponse. How does it compare with the framework presented here, and what set of questions is it designed to answer?
    (c) For a third example, describe the Oh and Scheuren (1983) quasi-randomization theory approach, and compare it to other frameworks.
    (d) Relate the various conditional randomization theory approaches to theory concerning anciliary statistics, as described, for example, in Cox and Hinkley (1974).

PROBLEMS

(e) Are the primary conclusions of Sections 2.9 and 2.10 altered if one of these other forms of randomization inference is used? Explain.

24. Assume a simple random sample. Prove that the expectation of $\bar{y}$ over its randomization distribution is $\bar{Y}$. Conclude that the expectation of Y_i^2 in the sample is the average value of Y_i^2 in the population. Prove that the expectation of s^2 is the population variance of Y_i.

25. Create a small population and draw many simple random samples from it. Produce stem-and-leaf displays of the randomization distributions of $\bar{y}$, s^2, and $z = (\bar{y} - \bar{Y})/[s(n^{-1} - N^{-1})^{1/2}]$.

26. Repeat Problem 25 with a population that will produce a nonnormal distribution for z; that is, create a population for which the standard inference for Y is inappropriate from the randomization perspective (hint: try skew Y).

27. Give examples in which the standard interval for $\bar{Y}$ is not appropriate from the Bayesian perspective because (a) the posterior distribution is not normal and then because (b) the posterior mean is not well approximated by $\bar{y}$.

28. Suppose the values of Y_i are approximately lognormally distributed. Describe a procedure for obtaining an approximate 95% interval estimate of $\bar{Y}$. Is the interval symmetric about a point estimate of $\bar{Y}$? (Hint: examine Rubin, 1983b).

29. Describe the type of interval estimates discussed in Section 2.8 and the different practical questions they address. Is the interval estimate constructed in Problem 28 one of the types and why or why not?

30. Suppose (c_1, c_2) is a 95% highest posterior density interval estimate for scalar positive Q. Is $(\log c_1, \log c_2)$ always a 95% highest posterior density interval estimate for $\log Q$ and why or why not?

31. Describe similarities and differences between the Bayesian significance level defined in Section 2.8 and the more traditional significance level associated with the test of a null hypothesis, for example, via the likelihood ratio or Wald test statistics. Similarly, relate Bayesian point estimates to traditional criteria for point estimation such as unbiasedness and minimum variance.

32. Prove that expressions (2.8.3) and (2.8.4) generate interval estimates of Q with coverage α under the appropriate assumptions on the distribution of Q.

33. Discuss Neyman's (1934) original rationale for using randomization-based inference.

34. Summarize the evaluation of interval estimates presented in Hansen, Madow, and Tepping (1983). Does it fit the paradigm for evaluating interval estimates described in Section 2.9, and why or why not?

35. Define calibration as used in Section 2.9. How does the definition compare with that presented in Dawid (1982)? Summarize the discussion in Dawid (1982).

36. Summarize the theoretical distinction between Bayesian and randomized-based coverage and why this distinction often may not be very important in applied practice.

37. Relate the ideas on calibration to concepts in statistical decision theory.

38. What property does Morris's (1983a) empirical Bayes confidence intervals possess?

39. Let X^- be a subset of the columns of X and Y^- be a subset of the columns of Y such that (i) the quantity to be estimated is a function only of X^- and Y^-:

$$Q(X,Y) = Q(X^-, Y^-)$$

and (ii) the sampling mechanism is such that inclusion/exclusion on Y^- is unconfounded given X^-:

$$\Pr(I^-|X,Y) = \Pr(I^-|X^-)$$

where I^- refers to the subset of columns of I corresponding to Y^-. Furthermore, (iii) let the interval C for Q depend only on X^-, Y^-_{obs}, R^-_{inc} and I^- (in the obvious notation):

$$C(X, Y_{obs}, R_{inc}, I) = C(X^-, Y^-_{obs}, R^-_{inc}, I^-).$$

(a) Show that the probability coverage of C given $(X^-, Y^-_{obs}, R^-_{inc}, I^-)$,

$$E[\delta(Q \in C)|X^-, Y^-_{obs}, R^-_{inc}, I^-],$$

depends explicitly only on the specification for $\Pr(X^-, Y^-, R^-)$. Now suppose that C has nominal level $1 - \alpha$ and is calibrated for all subsequences defined by $(X^-, Y^-_{obs}, R^-_{inc}, I^-)$:

$$E[\delta(Q \in C)|X^-, Y^-_{obs}, R^-_{inc}, I^-] = 1 - \alpha \text{ for all } X^-, Y^-_{obs}, R^-_{inc}, I^-.$$

(b) In general is C calibrated for all identifiable subsequences, that is, ones defined by (X, Y_{obs}, R_{inc}, I)?

(c) When is C calibrated for all identifiable sequences?

(d) Although C is generally not optimal under the full specification, can it still be a useful interval estimate of Q? For example, can it be useful to focus on a subset of variables (X^-, Y^-) and avoid specifying the distribution for all variables (X, Y)?

(e) Describe the consequences of violating (i).

(f) Describe the consequences of violating (ii).

(g) Summarize the implications of this problem regarding which variables should be the focus of modeling energy.

40. Suppose in addition to conditions (i), (ii), and (iii) of Problem 39, we have that (iv) the response mechanism is such that nonresponse on Y^- is ignorable given X^-, Y^-_{obs}:

$$\Pr(R^- | X, Y) = \Pr(R^- | X^-, Y^-_{obs}).$$

(a) Show that the probability coverage of C given $(X^-, Y^-_{obs}, R^-_{inc}, I^-)$ depends explicitly only on the specification for $\Pr(X^-, Y^-)$.

Now suppose that C has nominal level $1 - \alpha$ and is calibrated for all subsequences defined by $(X^-, Y^-_{obs}, R^-_{inc}, I^-)$.

(b) Provide modified answers to parts (b), (c), (d), (e), (f) of Problem 39 in the context of the additional assumption (iv).

(c) What is the effect of violating assumption (iv)?

(d) Summarize the implications of this problem regarding which variables should be the focus of modeling energy.

41. Describe a survey with many variables recorded and some nonresponse (e.g., the Current Population Survey) and a particular estimand Q of your choice. Discuss, in the light of Problems 39 and 40, which variables should receive most attention when deriving interval estimates of Q. Be explicit when describing the sampling mechanism and when making assumptions about the response mechanism.

42. Relate the content of Problems 39, 40, and 41 to the discussion in Rubin (1983c—expanded in Rubin, 1985) on Hansen, Madow, and Tepping (1983), and in particular to the idea of modeling using an "adequate" summary of covariates. Also see Sugden and Smith (1984), which refers to Rubin (1985).

43. Summarize the superpopulation approach described in Little (1982, 1983a, b) and any differences between it and the approach of this chapter.

44. Summarize evidence for the practical similarity of Bayesian and randomization-based inferences in the absence of nonresponse.

45. Summarize evidence for the practical similarity of Bayesian and randomization-based inference in the presence of nonresponse.

46. State the standard asymptotic results of Section 2.10 more carefully.

47. Consider the extensions of the standard asymptotic results to the case when the model used for analysis is wrong; summarize the results in Huber (1967).

48. Consider extensions of standard asymptotic results to finite population sampling; review the literature between Hajek (1960) and Mitra and Pathak (1984).

49. Comment on the plausibility of the major conclusion of Section 2.10 concerning nonresponse adjustments.

50. Compare and contrast the descriptions of Bayesian inference in surveys presented here and in Ericson (1969), Lindley (1972), and Little (1982 or 1983a, b).

CHAPTER 3

Underlying Bayesian Theory

3.1. INTRODUCTION AND SUMMARY OF REPEATED-IMPUTATION INFERENCES

The most straightforward justification for the use of multiple imputation arises from the Bayesian perspective, which not only provides a simple and general theoretical rationale but also provides prescriptions for how to create multiple imputations and analyze the resultant data in specific cases. In particular, if the multiple imputations are repetitions drawn to simulate a Bayesian posterior distribution of the missing values under a model, then appropriately combining analyses of each data set completed by imputation yields an approximately valid Bayesian inference under that model. The resulting inferences are called repeated-imputation inferences. A summary of these inferences serves as a useful introduction to the underlying theory, which is presented in the remaining sections of this chapter. Random-response randomization-based evaluations of these Bayesianly-derived repeated-imputation inferences appear in Chapter 4.

Notation

Let Q be the quantity of interest in the survey, for example, the mean Y in the population, $\bar{Y}$. Generally, Q is a k-dimensional row vector. Assume that with complete data, inferences for Q would be based on the statement that

$$(Q - \hat{Q}) \sim N(0, U) \qquad (3.1.1)$$

where $\hat{Q}$ is a statistic estimating Q, U is a statistic providing the variance ($k \times k$ covariance matrix generally) of $(Q - \hat{Q})$, and $N(0, U)$ is the k-variate normal distribution with mean 0 and variance U. In practice, it is prudent and common to choose Q to be some 1–1 function of the quantity of interest that makes the normal approximation (3.1.1) reasonable.

Suppose that under a specified Bayesian model, m sets of repeated imputations have been drawn and used to construct m completed data sets, where $\hat{Q}_{*1}, \ldots, \hat{Q}_{*m}$ and $U_{*1}, \ldots, U_{*m}$ are the values of the statistics, $\hat{Q}$ and U for each of these data sets. The theory of this chapter applies most directly when $\hat{Q}$ and U are *completed-data* statistics, which provide the mean and variance of Q given (X, Y_{inc}, R_{inc}, I) under the same Bayesian model as used to create the repeated imputations. Results in Section 3.6, however, show that in practice $\hat{Q}$ and U can often be *complete-data* statistics, which provide the mean and variance of Q given (X, Y_{inc}), and thus effectively ignore the observed value of R_{inc} and avoid the explicit conditioning on I when analyzing each completed data set [e.g., $\hat{Q} = \bar{y}$ and $U = s^2(n^{-1} - N^{-1})$ with n observations of scalar Y_i used to estimate $Q = \bar{Y}$]. In the following summary, we use the more familiar expression "complete-data statistics" for $\hat{Q}$ and U.

Combining the Repeated Complete-Data Estimates and Variances

The m repeated complete-data estimates and associated complete-data variances for Q under one model for nonresponse can be combined as follows. Let

$$\bar{Q}_m = \sum_{l=1}^{m} \hat{Q}_{*l}/m \qquad (3.1.2)$$

be the average of the m complete-data estimates,

$$\bar{U}_m = \sum_{l=1}^{m} U_{*l}/m \qquad (3.1.3)$$

be the average of the m complete-data variances, and

$$B_m = \sum_{l=1}^{m} (\hat{Q}_{*l} - \bar{Q}_m)^t (\hat{Q}_{*l} - \bar{Q}_m)/(m-1) \qquad (3.1.4)$$

be the variance between (among) the m complete-data estimates, where the superscript t indicates transpose when Q is a vector. The quantity

$$T_m = \bar{U}_m + (1 + m^{-1})B_m \qquad (3.1.5)$$

is the total variance of $(Q - \bar{Q}_m)$.

SUMMARY OF REPEATED-IMPUTATION INFERENCES

Scalar Q

Interval estimates and significance levels for scalar Q are formed using a t reference distribution with

$$\nu = (m-1)(1 + r_m^{-1})^2 \qquad (3.1.6)$$

degrees of freedom, where r_m is the relative increase in variance[†] due to nonresponse:

$$r_m = (1 + m^{-1}) B_m / \overline{U}_m. \qquad (3.1.7)$$

Thus, in the standard manner, a $100(1 - \alpha)\%$ interval estimate of Q is

$$\overline{Q}_m \pm t_\nu(\alpha/2) T_m^{1/2}, \qquad (3.1.8)$$

where $t_\nu(\alpha/2)$ is the upper $100\alpha/2$ percentage point of the student t distribution on ν degrees of freedom (e.g., if $\nu = \infty$ and $1 - \alpha = .95$, $t_\nu(\alpha/2) = 1.96$). Also, the significance level associated with the null value Q_0 is given by

$$\text{Prob}\{ F_{1,\nu} > (Q_0 - \overline{Q}_m)^2 / T_m \} \qquad (3.1.9)$$

where $F_{1,\nu}$ is an F random variable on one and ν degrees of freedom. The fraction of information about Q missing due to nonresponse is

$$\gamma_m = \frac{r_m + 2/(\nu + 3)}{r_m + 1}. \qquad (3.1.10)$$

Significance Levels Based on the Combined Estimates and Variances

When m is large relative to k (e.g., $m \geq 5k$), the significance level of the null value Q_0 of Q can be found by calculating

$$D_m = (Q_0 - \overline{Q}_m) T_m^{-1} (Q_0 - \overline{Q}_m)' / k, \qquad (3.1.11)$$

and letting the significance level be the probability that an F random variable on k and ν degrees of freedom is larger than D_m; ν is given by (3.1.6) with r_m generalized to be the average relative increase in variance

[†] More precisely, conditional variance given $B_\infty = B_m$.

due to nonresponse,

$$r_m = (1 + m^{-1})\text{Tr}(B_m \bar{U}_m^{-1})/k, \qquad (3.1.12)$$

where $\text{Tr}(A)$ is the sum of the diagonal elements in the $k \times k$ matrix A. When m is modest relative to k, a better test statistic is

$$\tilde{D}_m = (1 + r_m)^{-1}(Q_0 - \bar{Q}_m)\bar{U}_m^{-1}(Q_0 - \bar{Q}_m)^t/k, \qquad (3.1.13)$$

which is referred to the F distribution on k and $(k+1)\nu/2$ degrees of freedom. More work is needed to arrive at a single easily calculated and valid test statistic that combines the best features of D_m and $\tilde{D}_m$ when $k > 1$. With scalar Q, D_m and $\tilde{D}_m$ are identical, as are their reference distributions and give (3.1.9) for the significance level of Q_0.

Significance Levels Based on Repeated Complete-Data Significance Levels

A statistic asymptotically equivalent to $\tilde{D}_m$ can be computed from the m complete-data significance levels associated with Q_0 and the value of r_m. Specifically, let $d_{*1}, \ldots, d_{*m}$ be the m repeated values of the complete-data χ^2 statistics associated with the null value Q_0; that is, the significance level for the lth completed data set is the probability that a χ^2 random variable on k degrees of freedom is larger than d_{*l}. Then

$$\hat{D}_m = \frac{\dfrac{\bar{d}_m}{k} - \dfrac{m-1}{m+1}r_m}{1 + r_m}, \qquad (3.1.14)$$

where

$$\bar{d}_m = \sum_{l=1}^{m} d_{*l}/m = \text{the average repeated } \chi^2 \text{ statistic}, \qquad (3.1.15)$$

is referred to the F distribution on k and $(k+1)\nu/2$ degrees of freedom. The statistic $\hat{D}_m$ can be more useful than D_m or $\tilde{D}_m$ because it depends on the scalar χ^2 statistics $d_{*1}, \ldots, d_{*m}$ and the scalar r_m rather than on $k \times k$ matrices.

In some cases, only the estimated fraction of missing information for a particular component of Q may be known, perhaps because an interval estimate is to be created only for that component; using this in place of r_m in $\hat{D}_m$ would imply ν rather than $(k+1)\nu/2$ numerator degrees of freedom in the reference distribution for this estimated $\hat{D}_m$, assuming the fraction of

SUMMARY OF REPEATED-IMPUTATION INFERENCES

information missing on this component is typical of the other components. There does exist some information in the $d_{*1}, \ldots, d_{*m}$ about r_m, but it is not clear how best to use it to obtain p-values. Various suggestions are offered in Section 3.5. For example, a method of moments estimator of r_m, which can be used to provide estimates of both $\hat{D}_m$ and ν, is given by

$$\hat{r}_m = \frac{(1 + m^{-1})s_d^2}{2\bar{d}_m + [4\bar{d}_m^2 - 2ks_d^2]_+^{1/2}}, \tag{3.1.16}$$

where

$$s_d^2 = \sum_{l=1}^{m} (d_{*l} - \bar{d}_m)^2/(m-1). \tag{3.1.17}$$

The resultant values of (3.1.6) and (3.1.14), say $\hat{\nu}$ and $\hat{\hat{D}}_m$ are used to find the p-value for Q_0 by referring $\hat{\hat{D}}_m$ to an F distribution on k and $(1 + k^{-1})\hat{\nu}/2$ degrees of freedom.

Example 3.1. Inference for Regression Coefficients

Suppose that the object of inference in a large survey of an essentially infinite population is a two-component regression coefficient, $\beta = (\beta_1, \beta_2)$. Also, suppose that in the absence of missing data β would be estimated by the standard least-squares estimate b with associated variance–covariance matrix s^2V, where V^{-1} is the sum of cross-products matrix of the predictor variables and s^2 is the residual mean square in the least-squares regression. Thus, $Q = \beta$, $\hat{Q} = b$, $U = s^2V$ and the χ^2 statistic associated with $Q_0 = (0,0)$ is $bU^{-1}b'$. If the sample had not been large, an adjustment for using the normal reference distribution rather than a more appropriate t distribution would have been desirable. A simple adjustment due to Fisher (1935) in another context is to set U equal to $s^2V(f+3)/(f+1)$ where f is the degrees of freedom for the residual mean square. A better adjustment for the χ^2 statistic would be to look up the p-value for Q_0 using the $F_{k,f}$ reference distribution, and then find the corresponding χ_k^2 value, that is, the value d_0 such that $\text{Prob}\{F_{k,f} > b(s^2V)^{-1}b'\} = \text{Prob}\{\chi_k^2 > d_0\}$.

Now suppose that there are missing values, which have been multiply imputed to create five completed data sets under one model. The standard least-squares analysis is then performed on each completed data set. The result is: five repeated values of b, which we suppose have mean

$$\bar{Q}_m = (10, 20)$$

and variance–covariance matrix

$$B_m = \begin{bmatrix} 10 & 5 \\ 5 & 5 \end{bmatrix};$$

five values of s^2V, which we suppose have mean

$$\bar{U}_m = \begin{bmatrix} 13 & 14 \\ 14 & 44 \end{bmatrix};$$

and five χ^2 statistics testing $\beta = (0, 0)$ whose mean and variance we suppose are 15 and 10, respectively.

The relative increase in variance due to nonresponse is, from (3.1.7), 0.92 and 0.14 for β_1 and β_2, respectively, implying degrees of freedom in their t reference distributions equal to 17 and nearly 300, respectively. Thus from (3.1.8) the 95% interval for β_1 is given by $10 \pm t_{17}(.025)\sqrt{25}$ or $(-0.6, 20.6)$; similarly, the 95% interval for β_2 is given by $(6.1, 33.9)$. From (3.1.9), the p-value for $\beta_1 = 0$ is 0.06 and for $\beta_2 = 0$ is 0.005. From (3.1.10), the fraction of missing information for β_1 is 53% and for β_2 is 13%.

The significance level of the null value $\beta_0 = (0, 0)$ can be addressed by the statistics D_m, $\tilde{D}_m$, and $\hat{D}_m$. From (3.1.11) we have

$$D_m = (10, 20)\begin{bmatrix} 25 & 20 \\ 20 & 50 \end{bmatrix}^{-1}\begin{pmatrix} 10 \\ 20 \end{pmatrix}\Big/ 2 = 4.12.$$

Both $\tilde{D}_m$ and $\hat{D}_m$ depend on r_m, which from (3.1.12) equals 0.58. Thus from (3.1.13)

$$\tilde{D}_m = (1.58)^{-1}(10, 20)\begin{bmatrix} 13 & 14 \\ 14 & 44 \end{bmatrix}^{-1}\begin{pmatrix} 10 \\ 20 \end{pmatrix}\Big/ 2 = 3.37,$$

and from (3.1.14)

$$\hat{D}_m = (7.5 - 0.387)/1.58 = 4.50.$$

The degrees of freedom associated with D_m is $4(1.58/0.58)^2 = 30$, and with $\tilde{D}_m$ and $\hat{D}_m$ is $3/2 \times 30 = 45$. From tables of the $F_{2,30}$ distribution, the p-value associated with D_m is about 0.03, and from tables of the $F_{2,45}$ distribution, the p-values associated with $\tilde{D}_m$ and $\hat{D}_m$ are about 0.04 and 0.02, respectively.

Now suppose that only the five complete-data χ^2 test statistics were available for finding the significance level of the null value $\beta = (0, 0)$. Then

KEY RESULTS FOR ANALYSIS

by (3.1.16) and (3.1.17), a method of moments estimate of r_m is

$$\hat{r}_m = 12\left[30 + (4 \times 15^2 - 4 \times 10)^{1/2}\right]^{-1} \doteq 0.20,$$

and the associated estimated $\hat{D}_m$ is

$$\hat{D}_m = (7.5 - 0.133)/1.20 = 6.14,$$

which is referred to an $F_{2,.75\nu}$ distribution where $\hat{\nu} = 4(1.20/0.20)^2 \doteq 144$. The resulting p-value is approximately 0.003.

Finally, suppose that in addition to the five repeated χ^2 statistics, we have formed an interval estimate for one component of β and so have the relative increase in variance, r_m, for this component. When this component is β_1, $r_m = 0.92$ whereas when this component is β_2, $r_m = 0.14$. Using the β_1 value of r_m in (3.1.14) for $\hat{D}_m$ and in (3.1.6) for ν gives an estimated $\hat{D}_m$ equal to 4.36 with an associated $F_{k,\nu}$ reference distribution with $k = 2$ and $\nu = 17$, and a corresponding p-value equal to 0.030. In contrast, using the β_2 value of r_m gives an estimated $\hat{D}_m$ equal to 6.50 and a corresponding p-value of 0.002. As is clear from this example, p-values calculated directly from repeated significance levels can be quite sensitive to how r_m is estimated.

3.2. KEY RESULTS FOR ANALYSIS WHEN THE MULTIPLE IMPUTATIONS ARE REPEATED DRAWS FROM THE POSTERIOR DISTRIBUTION OF THE MISSING VALUES

Since the value of (X, Y_{obs}, R_{inc}, I) is observed, the actual posterior distribution of Q is

$$\Pr(Q|X, Y_{obs}, R_{inc}, I).$$

Suppose for the moment that we were told the values of Y_{mis}; that is, suppose that in addition to the observed values of X, Y_{obs}, R_{inc}, I, we fixed the values in Y_{mis}, as when they are imputed, to create a completed data set. The resultant conditional distribution of Q given both (X, Y_{obs}, R_{inc}, I) and the fixed value of Y_{mis} will be called the *completed-data posterior distribution of Q*:

$$\Pr(Q|X, Y_{inc}, R_{inc}, I), \tag{3.2.1}$$

where, as previously defined, $Y_{inc} = (Y_{obs}, Y_{mis})$.

Result 3.1. Averaging the Completed-Data Posterior Distribution of Q over the Posterior Distribution of Y_{mis} to Obtain the Actual Posterior Distribution of Q

The actual posterior distribution of $Q = Q(X, Y)$ equals the completed-data posterior distribution of Q averaged over the posterior distribution of the missing data, Y_{mis}:

$$\Pr(Q|X, Y_{obs}, R_{inc}, I) = \int \Pr(Q|X, Y_{inc}, R_{inc}, I)$$

$$\times \Pr(Y_{mis}|X, Y_{obs}, R_{inc}, I)\, dY_{mis}. \quad (3.2.2)$$

Result 3.1 is immediate because $Y_{inc} = (Y_{obs}, Y_{mis})$.

Result 3.1 can be applied to simulate the actual posterior distribution of Q using repeated draws from the posterior distribution of the missing values. Specifically, consider the posterior probability that Q is in some interval or region C, $\text{Prob}\{Q \in C | X, Y_{obs}, R_{inc}, I\}$, and let $Y_{mis}^{*1}, Y_{mis}^{*2}, \ldots, Y_{mis}^{*m}$ be m draws of Y_{mis} from its posterior distribution, $\Pr(Y_{mis}|X, Y_{obs}, R_{inc}, I)$. Let $Y_{inc}^{*1}, Y_{inc}^{*2}, \ldots, Y_{inc}^{*m}$ be the corresponding completed values of Y_{inc}, $Y_{inc}^{*l} = (Y_{obs}, Y_{mis}^{*l})$. Then

$$\text{Prob}\{Q \in C|X, Y_{obs}, R_{inc}, I\}$$

$$= \lim_{m \to \infty} \sum_{l=1}^{m} \text{Prob}\{Q \in C|X, Y_{inc}^{*l}, R_{inc}, I\}/m.$$

Example 3.2. The Normal Model Continued

Consider the normal setup of Examples 2.1 and 2.5 where the objective is to assess the posterior probability that the population mean $\overline{Y} = \sum_{1}^{N} Y_i/N$ is positive from a simple random sample of n values of Y_i. Thus $Q = \overline{Y}$ and the region of interest for Q is $C = (0, \infty)$. Using the notation of Example 2.1, the distribution of Q given Y_{inc} is t on $n - 1$ degrees of freedom with center $\bar{y}$ and scale $s(n^{-1} - N^{-1})^{1/2}$; thus

$$\text{Prob}\{\overline{Y} > 0|X, Y_{inc}\} = \int_{0}^{\infty} t[n - 1, \bar{y}, s^2(n^{-1} - N^{-1})]$$

in an obvious but inelegant notation. Suppose that due to unconfounded nonresponse, only n_1 of the n values in Y_{inc} are observed. Let $\bar{y}_1$ and s_1^2 be the sample mean and variance of the n_1 observed values in Y_{obs}. Because

KEY RESULTS FOR ANALYSIS

the sampling and response mechanisms are unconfounded and thus ignorable, it is easy to show from Result 2.3 that

$$\text{Prob}\{\bar{Y} > 0 | X, Y_{inc}, R_{inc}, I\} = \text{Prob}\{\bar{Y} > 0 | X, Y_{inc}\}$$

and

$$\text{Prob}\{\bar{Y} > 0 | X, Y_{obs}, R_{inc}, I\} = \text{Prob}\{\bar{Y} > 0 | X, Y_{obs}\}.$$

Furthermore, Lemma 2.1 implies that the m draws from the posterior distribution of Y_{mis} (i.e., the distribution of Y_{mis} given Y_{obs}) can be made as follows. For $l = 1, \ldots, m$, pass through the following three steps using independent draws for all random variables at each pass:

1. Draw a $\chi^2_{n_1-1}$ random variable, say x, and let

$$\sigma^2_* = s_1^2(m-1)/x;$$

2. Draw a $N(0,1)$ random variable, say z_0, and let

$$\mu_* = \bar{y}_1 + \sigma_* z_0 / n_1;$$

3. Draw $n - n_1$ independent $N(0,1)$ random variables, say z_i, $i \in mis$, and impute the missing components of Y_{inc} as

$$Y_{i*} = \mu_* + \sigma_* z_i, \quad i \in mis.$$

Each of the m draws of Y_{mis} from its posterior distribution creates a completed value of Y_{inc} from which $\text{Prob}\{\bar{Y} > 0 | X, Y_{inc}\}$ can be calculated as a t integral. The average value of the mt-integrals estimates $\text{Prob}\{\bar{Y} > 0 | X, Y_{obs}\}$, and for infinite m the average equals

$$\text{Prob}\{\bar{Y} > 0 | X, Y_{obs}\} = \int_0^\infty t\left[n_1 - 1, \bar{y}_1, s_1^2(n_1^{-1} - N^{-1})\right],$$

as can be shown by analysis.

The Posterior Cumulative Distribution Function of Q

In principal, $\text{Prob}\{Q > Q_0 | X, Y_{inc}, R_{inc}, I\}$ can be calculated for each completed data set and a sequence of values of Q_0, where $Q > Q_0$ means that each component of Q is larger than the corresponding component of Q_0. Then for each Q_0, $\text{Prob}\{Q > Q_0 | X, Y_{obs}, R_{inc}, I\}$ is calculated as the average value of $\text{Prob}\{Q > Q_0 | X, Y_{inc}, R_{inc}, I\}$ over the essentially infinite number of completed data sets. Although exceedingly tedious in application, if the sequence of values of Q_0 is large enough to cover the range of Q, this method essentially generates the posterior cumulative distribution function of Q. When only the posterior mean and variance–covariance matrix of Q are used to characterize the posterior distribution of Q, it is much easier to obtain the required summaries from repeated draws of Y_{mis}.

Result 3.2. Posterior Mean and Variance of Q

Suppose that the completed-data posterior distribution of $Q = Q(X, Y)$ has mean $\hat{Q} = \hat{Q}(X, Y_{inc}, R_{inc}, I)$ and variance (or variance–covariance matrix if $k > 1$) $U = U(X, Y_{inc}, R_{inc}, I)$. Then the posterior distribution of Q has mean given by

$$E(Q|X, Y_{obs}, R_{inc}, I) = E(\hat{Q}|X, Y_{obs}, R_{inc}, I), \qquad (3.2.3)$$

and variance given by

$$V(Q|X, Y_{obs}, R_{inc}, I) = E(U|X, Y_{obs}, R_{inc}, I) + V(\hat{Q}|X, Y_{obs}, R_{inc}, I). \qquad (3.2.4)$$

Result 3.2 is immediate from the rules for finding unconditional moments from conditional moments, given in Section 2.3, and the definitions of $\hat{Q}$ and U. That is,

$$E(Q|X, Y_{obs}, R_{inc}, I) = E\big[E(Q|X, Y_{obs}, Y_{mis}, R_{inc}, I)|X, Y_{obs}, R_{inc}, I\big]$$

and

$$V(Q|X, Y_{obs}, R_{inc}, I) = V\big[E(Q|X, Y_{obs}, Y_{mis}, R_{inc}, I)|X, Y_{obs}, R_{inc}, I\big]$$
$$+ E\big[V(Q|X, Y_{obs}, Y_{mis}, R_{inc}, I)|X, Y_{obs}, R_{inc}, I\big],$$

where, since $Y_{inc} = (Y_{obs}, Y_{mis})$,

$$E(Q|X, Y_{obs}, Y_{mis}, R_{inc}, I) = \hat{Q}$$

and

$$V(Q|X, Y_{obs}, Y_{mis}, R_{inc}, I) = U,$$

and it follows that (3.2.3) and (3.2.4) hold.

Result 3.2 is important because it suggests a simple numerical procedure for simulating the posterior mean and variance of $Q(X, Y)$.

Simulating the Posterior Mean and Variance of Q

Suppose that there exist m independent draws from the posterior distribution of Y_{mis}, where for simplicity m is essentially infinite. These can be used as imputed values to create m completed data sets and thereby m values of the statistics $\hat{Q}$ and U. These m repeated values of $\hat{Q}$ and U, say $\hat{Q}_{*1}, \ldots, \hat{Q}_{*m}$ and $U_{*1}, \ldots, U_{*m}$, simulate aspects of the posterior distribution of $\hat{Q}$ and U. Specifically, since m is essentially infinite, (i) the average $\hat{Q}_{*l}$ gives the posterior mean of $\hat{Q}$.

$$\bar{Q}_{\infty} \equiv \lim_{m \to \infty} \sum_{l=1}^{m} \hat{Q}_{*l}/m = E(\hat{Q}|X, Y_{obs}, R_{inc}, I), \quad (3.2.5)$$

(ii) the average U_{*l} gives the posterior mean of U,

$$\bar{U}_{\infty} \equiv \lim_{m \to \infty} \sum_{l=1}^{m} U_{*l}/m = E(U|X, Y_{obs}, R_{inc}, I), \quad (3.2.6)$$

and (iii) the variance among the $\hat{Q}_{*l}$ gives the posterior variance of $\hat{Q}$,

$$B_{\infty} \equiv \lim_{m \to \infty} \sum_{l=1}^{m} (\hat{Q}_{*l} - \bar{Q}_{\infty})^t (\hat{Q}_{*l} - \bar{Q}_{\infty})/m = V(\hat{Q}|X, Y_{obs}, R_{inc}, I). \quad (3.2.7)$$

From (3.2.3) and (3.2.5)

$$E(Q|X, Y_{obs}, R_{inc}, I) = \bar{Q}_{\infty}, \quad (3.2.8)$$

and from (3.2.4), (3.2.6), and (3.2.7)

$$V(Q|X, Y_{obs}, R_{inc}, I) = T_{\infty} \quad (3.2.9)$$

where

$$T_{\infty} = \bar{U}_{\infty} + B_{\infty}. \quad (3.2.10)$$

Missing and Observed Information with Infinite m

It is conceptually useful to relate the variance–covariance matrices in equation (3.2.10) to concepts of missing and observed information. Define the information that is observed to be T_{∞}^{-1} and the expected total information if Y_{mis} were also observed to be $E(U|X, Y_{obs}, R_{inc}, I) = \overline{U}_\infty^{-1}$. Then the missing information is $\overline{U}_\infty^{-1} - T_\infty^{-1}$, and the fraction of information that is missing can be written as

$$\gamma_\infty = \left(\overline{U}_\infty^{-1}\right)^{-1/2} \left(\overline{U}_\infty^{-1} - T_\infty^{-1}\right) \left(\overline{U}_\infty^{-1}\right)^{-1/2} \qquad (3.2.11)$$

using the symmetric form of the matrix generalization of dividing the missing information by the expected total information. Simple matrix algebra gives an alternative expression for γ_∞:

$$\gamma_\infty = I - \overline{U}_\infty^{1/2} T_\infty^{-1} \overline{U}_\infty^{1/2} = I - T_\infty^{-1/2} \overline{U}_\infty T_\infty^{-1/2}$$

$$= T_\infty^{-1/2} B_\infty T_\infty^{-1/2}. \qquad (3.2.12)$$

The fraction of information missing due to nonresponse is thus given by the eigenvalues of B_∞ with respect to T_∞, where B_∞ is the increase in the posterior variance of Q due to nonresponse, and T_∞ is the total posterior variance of Q. With scalar Q, $\gamma_\infty = r_\infty/(r_\infty + 1)$, where r_∞ is defined by (3.1.7) to be $B_\infty/\overline{U}_\infty$.

Inference for Q from Repeated Completed-Data Means and Variances

Using (3.2.8) and (3.2.9) and the usual approximation of the posterior distribution as normal, motivated in Section 2.10, we are led to the statement that

$$(Q - \overline{Q}_\infty | X, Y_{obs}, R_{inc}, I) \sim N(0, T_\infty). \qquad (3.2.13)$$

Interval estimates for Q and significance levels for null values of Q follow from (3.2.13) in the standard manner. Thus with scalar Q,

$$\overline{Q}_\infty \pm 1.96 T_\infty^{1/2} \qquad (3.2.14)$$

is a 95% interval estimate of Q. Similarly, from expression (2.8.7) the significance level associated with the null value of Q, Q_0, is

$$p\text{-value}(Q_0|X, Y_{obs}, R_{inc}, I) = \text{Prob}\{\chi_k^2 > kD_\infty\} \qquad (3.2.15)$$

where χ_k^2 is a χ^2 random variable on k degrees of freedom (k the dimension of Q) and D_∞ is the observed value

$$D_\infty = (Q_0 - \overline{Q}_\infty) T_\infty^{-1} (Q_0 - \overline{Q}_\infty)'/k. \qquad (3.2.16)$$

INFERENCE FOR SCALAR ESTIMANDS

Example 3.3. Example 3.2 Continued

Suppose $Q = \overline{Y}$, $\hat{Q} = \bar{y}$, and $U = s^2(n^{-1} - N^{-1})$, in the setup of Example 3.2, where an essentially infinite number of repeated multiple imputations are created using the normal model of Example 3.2. Then with $m \to \infty$,

$$\overline{Q}_\infty = \sum_{l=1}^{m} \bar{y}_{*l}/m$$

and

$$T_\infty = \sum_{l=1}^{m} s_{l*}^2(n^{-1} - N^{-1})/m + \sum_{l=1}^{m} (\bar{y}_{*l} - \overline{Q}_\infty)^2/m.$$

The 95% interval for $\overline{Y}$ is given by (3.2.14), and $1 - \Phi(-\overline{Q}_\infty/T_\infty^{1/2})$ is the posterior probability that $\overline{Y}$ is positive. This probability can also be found by calculating D_∞ in (3.2.15) with $Q_0 = 0$ and $k = 1$, and finding the probability that a χ^2 random variable on one degree of freedom is greater than the observed value of D_∞.

3.3. INFERENCE FOR SCALAR ESTIMANDS FROM A MODEST NUMBER OF REPEATED COMPLETED-DATA MEANS AND VARIANCES

Thus far, we have implicitly assumed that the actual posterior distribution of Y_{mis} could be simulated perfectly in the sense that inferences have been based on an infinite number of completed data sets. Of course, in practice a modest number of imputations must be made, and therefore if one inference is desired under an imputation model, we need to modify the theory of this chapter to be appropriate with finite m. The essential idea is that we must condition on the m values of the repeated completed-data summary statistics we observe rather than on the infinite number we would ideally observe. The required conditioning is especially straightforward in the very common situation with scalar Q where completed-data posterior distributions are assumed to be normal and each is summarized by its mean and variance. The following development derives an approximating t distribution for Q given m completed-data means and variances. Estimates and significance levels can then be obtained using the t and F reference distributions. Extensions to multicomponent Q are outlined in Sections 3.4 and 3.5.

The Plan of Attack

Suppose m draws of Y_{mis} are taken from its posterior distribution, $\Pr(Y_{mis}|X, Y_{obs}, R_{inc}, I)$, and let $\mathbf{S}_m = \{\hat{Q}_{*l}, U_{*l}, l = 1, \ldots, m\}$ be the set of associated completed-data statistics

$$\hat{Q} = E(Q|X, Y_{inc}, R_{inc}, I) \quad \text{and} \quad U = V(Q|X, Y_{inc}, R_{inc}, I)$$

evaluated on each of the m completed data sets. We wish to approximate the conditional distribution of Q given $\mathbf{S}_m$.

If $\mathbf{S}_\infty$ were observed, following (3.2.13) we would approximate the posterior distribution of Q as normal with mean $\overline{Q}_\infty$ and variance $T_\infty = \overline{U}_\infty + B_\infty$:

$$(Q|X, Y_{obs}, R_{inc}, I) \sim N(\overline{Q}_\infty, \overline{U}_\infty + B_\infty),$$

where the conditioning on (X, Y_{obs}, R_{inc}, I) can be replaced by conditioning on $\mathbf{S}_\infty$ or simply on $(\overline{Q}_\infty, \overline{U}_\infty, B_\infty)$:

$$(Q|\overline{Q}_\infty, \overline{U}_\infty, B_\infty) \sim N(\overline{Q}_\infty, \overline{U}_\infty + B_\infty). \quad (3.3.1)$$

Our problem is to approximate the conditional distribution of Q given $\mathbf{S}_m$ rather than $\mathbf{S}_\infty$. To go from (3.3.1), which is the same as the conditional distribution of Q given $(\overline{Q}_\infty, \overline{U}_\infty, B_\infty)$ and $\mathbf{S}_m$, to the conditional distribution of Q given $\mathbf{S}_m$, we approximate the conditional distribution of $(\overline{Q}_\infty, \overline{U}_\infty, B_\infty)$ given $\mathbf{S}_m$ and average (3.3.1) over this conditional distribution.

The Sampling Distribution of $\mathbf{S}_m$ Given (X, Y_{obs}, R_{inc})

The first step is to consider the sampling distribution of $\mathbf{S}_m$ given (X, Y_{obs}, R_{inc}), where we recall from (3.2.5)–(3.2.7) that

$$\overline{Q}_\infty = E(\hat{Q}|X, Y_{obs}, R_{inc}, I),$$

$$\overline{U}_\infty = E(U|X, Y_{obs}, R_{inc}, I),$$

and

$$B_\infty = V(\hat{Q}|X, Y_{obs}, R_{inc}, I).$$

By construction, the $(\hat{Q}_{*l}, U_{*l})$ in $\mathbf{S}_m$ are m i.i.d. draws of the posterior mean and variance of Q, that is, of $\hat{Q}$ and U from their joint

INFERENCE FOR SCALAR ESTIMANDS

sampling distribution, where the distribution of the "data," Y_{mis}, is $\Pr(Y_{mis}|X, Y_{obs}, R_{inc}, I)$. Asymptotic theory of posterior means and variances summarized in Section 2.10 suggests that under rather general conditions with large samples and populations, the sampling distribution of the posterior mean tends toward normality, and the sampling distribution of the posterior variance tends to have lower-order variability than that of the posterior mean. Thus, with large data sets it may be reasonable to suppose that given (X, Y_{obs}, R_{inc}, I), $\mathbf{S}_m$ consists of m i.i.d. draws from

$$(\hat{Q}_{*l}|X, Y_{obs}, R_{inc}, I) \sim N(\overline{Q}_\infty, B_\infty) \qquad (3.3.2)$$

$$(U_{*l}|X, Y_{obs}, R_{inc}, I) \sim (\overline{U}_\infty, \ll B_\infty) \qquad (3.3.3)$$

using the notation in Section 2.10.

The Conditional Distribution of $(\overline{Q}_\infty, \overline{U}_\infty)$ Given $\mathbf{S}_m$ and B_∞

Suppose we accept the approximations given by (3.3.2) and (3.3.3) for the conditional distribution of the $(\hat{Q}_{*l}, U_{*l})$ given $(\overline{Q}_\infty, \overline{U}_\infty, B_\infty)$. Then an approximating conditional distribution for $(\overline{Q}_\infty, \overline{U}_\infty)$ given $\mathbf{S}_m$ and B_∞ is easy to derive. First, by (3.3.3), for any relatively diffuse prior distribution on $\overline{U}_\infty$,

$$(\overline{U}_\infty|\mathbf{S}_m, B_\infty) \sim (\overline{U}_m, \ll B_\infty/m) \qquad (3.3.4)$$

where

$$\overline{U}_m = \sum_{l=1}^{m} U_{*l}/m.$$

Next, if the prior distribution of $\overline{Q}_\infty$ given B_∞ is proportional to a constant, then because of the normal sampling distribution of the $\hat{Q}_{*l}$ in (3.3.2), the conditional distribution of $\overline{Q}_\infty$ given $\mathbf{S}_m$ and B_∞ is normal:

$$(\overline{Q}_\infty|\mathbf{S}_m, B_\infty) \sim N(\overline{Q}_m, B_\infty/m) \qquad (3.3.5)$$

where

$$\overline{Q}_m = \sum_{l=1}^{m} \hat{Q}_{*l}/m.$$

The Conditional Distribution of Q Given $\mathbf{S}_m$ and B_∞

Having now approximated the conditional distribution of $(\overline{Q}_\infty, \overline{U}_\infty)$ given $\mathbf{S}_m$ and B_∞, we can apply it to (3.3.1), the conditional distribution of Q

given $(\overline{Q}_\infty, \overline{U}_\infty, \overline{B}_\infty)$ and implicitly S_m, to obtain the conditional distribution of Q given S_m and B_∞. Expression (3.3.4) implies that in (3.3.1) $\overline{U}_\infty$ can be replaced by $\overline{U}_m$ to give

$$(Q|S_m, \overline{Q}_\infty, B_\infty) \sim N(\overline{Q}_\infty, \overline{U}_m + B_\infty). \qquad (3.3.6)$$

Expression (3.3.5) applied to (3.3.6) gives

$$(Q|S_m, B_\infty) \sim N(\overline{Q}_m, \overline{U}_m + (1 + m^{-1})B_\infty). \qquad (3.3.7)$$

The Conditional Distribution of B_∞ Given S_m

Thus far our analysis has been conditional on B_∞ as well as S_m. We now address the problem of integrating over B_∞. Assume that the prior distribution on $\log(B_\infty)$ is proportional to a constant. Under the asymptotic sampling distributions given by (3.3.2) and (3.3.3), the conditional distribution of B_∞ given S_m is then proportional to an inverted χ^2 random variable on $m - 1$ degrees of freedom, as implied by Lemma 2.1:

$$((m-1)B_m B_\infty^{-1}|S_m) \sim \chi^2_{m-1}, \qquad (3.3.8)$$

where

$$B_m = \sum_{l=1}^m (\hat{Q}_{*l} - \overline{Q}_m)^2/(m-1)$$

is the standard unbiased estimate of the variance, B_∞.

The Conditional Distribution of $\overline{U}_m + (1 + m^{-1})B_\infty$ Given S_m

From (3.3.7), the conditional variance of Q given B_∞ and S_m is $\overline{U}_m + (1 + m^{-1})B_\infty$. If given S_m, $\overline{U}_m + (1 + m^{-1})B_\infty$ were proportional to an inverted χ^2 random variable, (3.3.7) would imply that Q given S_m had a t distribution. Unfortunately, even though B_∞ is proportional to an inverted χ^2 random variable, a constant plus an inverted χ^2 is not distributed as an inverted χ^2, and in fact Q given S_m has a Behrens–Fisher distribution rather than a t (see Box and Tiao, 1973, p. 106). Nevertheless this distribution can be well approximated by a t. The approximation presented here first approximates the conditional distribution of $\overline{U}_m + (1 + m^{-1})B_\infty$ given S_m as proportional to an inverted χ^2, and then derives the corresponding t.

INFERENCE FOR SCALAR ESTIMANDS

Specifically, we adopt the following approximation:

$$\left(\nu T_m [\overline{U}_m + (1 + m^{-1})B_\infty]^{-1} \big| S_m\right) \sim \chi_\nu^2 \tag{3.3.9}$$

where T_m is the estimated total variance

$$T_m = \overline{U}_m + (1 + m^{-1})B_m, \tag{3.3.10}$$

and ν is the degrees of freedom

$$\nu = (m - 1)(1 + r_m^{-1})^2 \tag{3.3.11}$$

where

$$r_m = (1 + m^{-1})B_m/\overline{U}_m. \tag{3.3.12}$$

From (3.3.7) r_m is the relative increase in conditional variance due to nonresponse, given $B_\infty = B_m$. The basis for (3.3.9) is Approximation 3.1.

Approximation 3.1 Relevant to the Behrens-Fisher Distribution

Suppose x^{-1} is distributed as a mean-square random variable on f degrees of freedom (i.e., fx^{-1} is distributed as χ_f^2). Then $(1 + a)/(1 + ax)$ is distributed approximately as a mean-square random variable with degrees of freedom equal to $f(1 + a^{-1})^2$.

The basic idea of this approximation is to fit the first two moments—approximately. Since a mean-square random variable has mean one and variance two divided by its degrees of freedom, we show that

$$E\left(\frac{1 + a}{1 + ax}\right) \doteq 1, \tag{3.3.13}$$

and

$$V\left(\frac{1 + a}{1 + ax}\right) \doteq 2\left[f\left(\frac{1 + a}{a}\right)^2\right]^{-1}. \tag{3.3.14}$$

Since $(1 + a)/(1 + ax)$ is nonlinear in the mean-square random variable x^{-1}, we use a first term Taylor series expansion in x^{-1} about its mean:

$$\frac{1 + a}{1 + ax} \doteq 1 + \frac{a}{1 + a}(x^{-1} - 1). \tag{3.3.15}$$

The expectation of the right-hand side of (3.3.15) is 1, as in (3.3.13), and the variance of the right-hand side of (3.3.15) is $(1 + a^{-1})^2(2/f)$, which equals the right-hand side of (3.3.14).

Applying Approximation 3.1 to Obtain (3.3.9)

The quantity $\overline{U}_m + (1 + m^{-1})B_\infty$ in (3.3.9) can be written as

$$\overline{U}_m(1 + r_m B_\infty/B_m)$$

where $(B_\infty/B_m)^{-1}$ given $\mathbf{S}_m$ is distributed as a mean-square random variable on $m - 1$ degrees of freedom. Hence, by Approximation 3.1, the quantity

$$\frac{\overline{U}_m(1 + r_m)}{\overline{U}_m(1 + r_m B_\infty/B_m)} = \frac{T_m}{\overline{U}_m + (1 + m^{-1})B_\infty}$$

is distributed approximately as a mean-square random variable on ν degrees of freedom, where ν is given by (3.3.11), from which (3.3.9) is immediate.

The Approximating t Reference Distribution for Scalar Q

The resulting t approximation for the distribution of Q given the m completed moments is

$$(Q|\mathbf{S}_m) \sim t_\nu(\overline{Q}_m, T_m) \tag{3.3.16}$$

where ν is given by (3.3.11) and T_m is given by (3.3.10). Consequently, interval estimates are based on (3.3.16), and significance tests for null values, Q_0, are obtained by referring $(\overline{Q}_m - Q_0)^2/T_m$ to an F distribution on one and ν degrees of freedom.

Example 3.4. Example 3.3 Continued

Suppose $Q = \overline{Y}$, $\hat{Q} = \bar{y}$, and $U = s^2(n^{-1} - N^{-1})$ in the setup of Example 3.3, where m repeated imputations are created using the normal model of

INFERENCE FOR SCALAR ESTIMANDS

Example 3.2. Then

$$\bar{Q}_m = \sum_{l=1}^{m} \bar{y}_{*l}/m,$$

$$\bar{U}_m = \sum_{l=1}^{m} s_{*l}^2 (n^{-1} - N^{-1})/m$$

and

$$B_m = \sum_{l=1}^{m} (\bar{y}_{*l} - \bar{Q}_m)^2/(m-1),$$

where T_m is given by (3.3.10) and ν is given by (3.3.11). The $1 - \alpha$ interval estimate for $\bar{Y}$ is given by

$$\bar{Q}_m \pm t_\nu(\alpha/2) T_m^{1/2}$$

and the significance level of the null value Q_0 is

$$\text{Prob}\{F_{1,\nu} > (\bar{Q}_m - Q_0)^2/T_m\}.$$

Fraction of Information Missing Due to Nonresponse

The average second derivative of the logarithm of the t posterior distribution for Q given by (3.3.16) equals $-(\nu + 1)(\nu + 3)^{-1} T_m^{-1}$. Hence, using the definition for information proposed in Fisher (1935), the information about Q in this posterior distribution is $(\nu + 1)(\nu + 3)^{-1} T_m^{-1}$. If there had been complete response, the posterior distribution of Q would have been normal with expected second derivative of the log posterior equal to $-\bar{U}_\infty^{-1}$ or $-\bar{U}_m^{-1}$ in our asymptotic analysis. Thus, the fraction of information missing due to nonresponse is

$$\gamma_m = \left[\bar{U}_m^{-1} - (\nu + 1)(\nu + 3)^{-1} T_m^{-1}\right]/\bar{U}_m^{-1},$$

or from (3.3.10) and (3.3.12),

$$\gamma_m = \frac{r_m + 2/(\nu + 3)}{r_m + 1} \qquad (3.3.17)$$

where ν is given as a function of m and r_m by (3.3.11). When $m = \infty$, $\gamma_\infty = r_\infty/(r_\infty + 1)$ in agreement with (3.2.12).

3.4. SIGNIFICANCE LEVELS FOR MULTICOMPONENT ESTIMANDS FROM A MODEST NUMBER OF REPEATED COMPLETED-DATA MEANS AND VARIANCE–COVARIANCE MATRICES

Section 3.3 dealt with inference from a finite number of imputations but restricted attention to the case of a scalar estimand. Extending that analysis to k-component Q is straightforward up to a point. That point involves the integration over the distribution of B_∞ given S_m. Consequently, we first indicate simple extensions given B_∞ and then consider additional issues created by the need to integrate over B_∞. The plan of attack is identical to that in Section 3.3, and in fact equations (3.3.1)–(3.3.7) hold for vector Q without modification assuming the prior distribution on $\overline{Q}_\infty$ is proportional to a constant.

The Conditional Distribution of Q Given S_m and B_∞

Specifically, the distribution of k-component Q given (S_m, B_∞) is the k-variate normal with mean $\overline{Q}_m$ and variance–covariance matrix $\overline{U}_m + (1 + m^{-1})B_\infty$:

$$(Q|S_m, B_\infty) \sim N(\overline{Q}_m, \overline{U}_m + (1 + m^{-1})B_\infty). \qquad (3.4.1)$$

If B_∞ were known and m completed-data moments were available, (3.4.1) would be a normal reference distribution for interval estimates of Q. Furthermore, (3.4.1) implies a χ^2 reference distribution on k degrees of freedom for obtaining the significance level associated with a null value of Q, Q_0. Specifically, the p-value associated with Q_0 given (S_m, B_∞) is

p-value$(Q_0|S_m, B_\infty)$

$$= \text{Prob}\left\{\chi_k^2 > (Q_0 - \overline{Q}_m)[\overline{U}_m + (1 + m^{-1})B_\infty]^{-1}(Q_0 - \overline{Q}_m)'\right\}. \qquad (3.4.2)$$

The Bayesian p-Value for a Null Value Q_0 Given S_m: General Expression

From (3.4.1), given S_m and B_∞, the correct Bayesian p-value for Q_0 is provided by (3.4.2). When B_∞ is not known, the correct Bayesian p-value is the average value of (3.4.2) with respect to the conditional distribution of B_∞ given S_m:

$$p\text{-value}(Q_0|S_m) = \int \text{Prob}\{\chi_k^2 > (Q_0 - \overline{Q}_m)[\overline{U}_m + (1 + m^{-1})B_\infty]^{-1}$$

$$\times (Q_0 - \overline{Q}_m)^t\} \Pr(B_\infty|S_m) \, dB_\infty. \quad (3.4.3)$$

There are thus two problems in finding the p-value: first, finding the distribution of B_∞ given S_m, and second, performing the integration called for in (3.4.3). The distribution of B_∞ given S_m follows from the distribution of B_m given B_∞, which is Wishart, and the prior distribution of B_∞. Evaluating (3.4.3) can be quite complicated; even the case of scalar Q exposes several relevant issues.

The Bayesian p-Value Given S_m with Scalar Q

In the case of scalar Q, (3.4.3) can be written as

$$p\text{-value}(Q_0|S_m; k = 1) = \int \text{Prob}\{\chi_1^2 > [1 + (1 + m^{-1})r_\infty]^{-1}$$

$$\times (Q_0 - \overline{Q}_m)^2 \overline{U}_m^{-1}\} \Pr(r_\infty|S_m) \, dr_\infty \quad (3.4.4)$$

where $B_\infty = r_\infty \overline{U}_\infty$. Section 3.3 showed that with a locally uniform prior distribution on $\log(B_\infty)$, the distribution of B_∞ given S_m is $(m-1)B_m$ times an inverted χ^2 random variable on $m-1$ degrees of freedom; hence $(1 + m^{-1})r_\infty$ given S_m is $(m-1)r_m$ times an inverted χ^2 random variable on $m-1$ degrees of freedom, where $(1 + m^{-1})B_m = r_m\overline{U}_m = r_m\overline{U}_\infty$ in our asymptotic analysis. Hence,

$$p\text{-value}(Q_0|S_m; k = 1)$$

$$= \text{Prob}\left\{\chi_1^2 > \left[1 + \frac{(m-1)}{\chi_{m-1}^2}r_m\right]^{-1} (Q_0 - \overline{Q}_m)^2 \overline{U}_m^{-1}\right\} \quad (3.4.5)$$

where χ_1^2 and χ_{m-1}^2 are independent χ^2 random variables. With observed values of the statistics $(r_m, \overline{Q}_m, \overline{U}_m)$, (3.4.5) can be evaluated by numerical integration or Monte Carlo simulation—draw many independent χ_1^2 and χ_{m-1}^2 random variables and find the proportion of times the inequality in (3.4.5) is satisfied. Nevertheless, a simple closed-form approximation to (3.4.5) is clearly useful in many cases.

The Bayesian p-Value Given S_m with Scalar Q—Closed-Form Approximation

In Section 3.3 the distribution of $\overline{U}_m + (1 + m^{-1})B_\infty$ given S_m was approximated as νT_m times an inverted χ^2 random variable on ν degrees of freedom where ν is given by (3.3.11). Using the same approximation here implies that $[1 + (1 + m^{-1})r_\infty]$ in (3.4.4) can be approximated as $\nu T_m \overline{U}_m^{-1}$ times an inverted χ^2 random variable on ν degrees of freedom. Hence, (3.4.4) can be approximated as

$$p\text{-value}(Q_0|S_m; k = 1) \doteq \text{Prob}\{\chi_1^2 > (\chi_\nu^2/\nu)(Q_0 - \overline{Q}_m)^2 T_m^{-1}\}$$

or

$$p\text{-value}(Q_0|S_m; k = 1) \doteq \text{Prob}\{F_{1,\nu} > (Q_0 - \overline{Q}_m)^2 T_m^{-1}\}, \quad (3.4.6)$$

where $F_{1,\nu}$ is an F random variable on 1 and ν degrees of freedom; (3.4.6) is the same p-value as suggested following (3.3.16) and illustrated in Example 3.4.

This analysis with scalar Q can be generalized to multicomponent Q in special cases when B_∞ is proportional to T_∞.

p-Values with B_∞ a Priori Proportional to T_∞

Generally the analysis is much less tidy with k-component Q than with scalar Q. An exception occurs when *a priori* B_∞ is proportional to T_∞ and thus to $\overline{U}_\infty$, that is, when the fractions of information missing for the k individual components of Q are all the same. Since $\overline{U}_\infty$ essentially equals $\overline{U}_m$ in our asymptotic analysis, the only quantity that needs to be estimated then is the constant of proportionality r_∞ where $B_\infty = r_\infty \overline{U}_\infty$. In that case (3.4.3) becomes

$$p\text{-value}(Q_0|S_m; B_\infty \propto \overline{U}_\infty)$$

$$= \int \text{Prob}\{\chi_k^2 > [1 + (1 + m^{-1})r_\infty]^{-1}(Q_0 - \overline{Q}_m)\overline{U}_m^{-1}(Q_0 - \overline{Q}_m)'\}$$

$$\times \Pr(r_\infty|S_m)\, dr_\infty. \quad (3.4.7)$$

When the prior distribution on $\log(r_\infty)$ is proportional to a constant, it is

SIGNIFICANCE LEVELS FOR MULTICOMPONENT ESTIMANDS

easy to show using standard multivariate normal theory that the distribution of $(1 + m^{-1})r_\infty$ given $\mathbf{S}_m$ is $k(m-1)r_m$ times an inverted χ^2 random variable on $k(m-1)$ degrees of freedom, where

$$r_m = (1 + m^{-1})\text{Tr}(B_m \bar{U}_m^{-1})/k; \qquad (3.4.8)$$

the basic idea is that after an appropriate linear transformation defined by $\bar{U}_m^{-1/2}$, each component of Q provides $m-1$ independent degrees of freedom for estimating r_∞, just as with scalar Q, and when these are added together they provide $k(m-1)$ degrees of freedom for estimating r_∞. Hence,

$$p\text{-value}(Q_0 | \mathbf{S}_m; B_\infty \propto \bar{U}_\infty) = \text{Prob}\left\{\chi_k^2 > \left[1 + \frac{k(m-1)}{\chi_{k(m-1)}^2}r_m\right]^{-1}\right.$$
$$\left. \times (Q_0 - \bar{Q}_m)\bar{U}_m^{-1}(Q_0 - \bar{Q}_m)^t\right\}. \qquad (3.4.9)$$

With observed values of the statistics $(r_m, \bar{Q}_m, \bar{U}_m)$, (3.4.9) can be evaluated by numerical integration or Monte Carlo simulation: draw many pairs of independent χ_k^2 and $\chi_{k(m-1)}^2$ random variables and find the proportion of times the inequality in braces in (3.4.9) is satisfied. Furthermore, an approximation analogous to (3.4.6) used with $k=1$ can be derived to provide a closed-form expression.

p-Values with B_∞ a Priori Proportional to T_∞—Closed-Form Approximation

Approximation 3.1 can be used to show that the random variable in square brackets in (3.4.9) can be approximated as proportional to an inverted χ^2 random variable. Specifically,

$$(1 + r_m)\left[1 + \frac{k(m-1)}{\chi_{k(m-1)}^2}r_m\right]^{-1}$$

can be approximated as a mean-square random variable on $k\nu$ degrees of freedom where ν is given by (3.3.11). It follows that (3.4.9) can be approximated as

$$p\text{-value}(Q_0 | \mathbf{S}_m; B_\infty \propto \bar{U}_\infty)$$
$$\doteq \text{Prob}\left\{\chi_k^2 > \chi_{k\nu}^2(1+r_m)^{-1}(k\nu)^{-1}(Q_0 - \bar{Q}_m)\bar{U}_m^{-1}(Q_0 - \bar{Q}_m)^t\right\}$$

where χ_k^2 and $\chi_{k\nu}^2$ are independent χ^2 random variables. Equivalently,

$$p\text{-value}(Q_0|S_m; B_\infty \propto U_\infty) \doteq \text{Prob}\{F_{k,k\nu} > \tilde{D}_m\} \qquad (3.4.10)$$

where

$$\tilde{D}_m = (1 + r_m)^{-1}(Q_0 - \overline{Q}_m)\overline{U}_m^{-1}(Q_0 - \overline{Q}_m)^t/k. \qquad (3.4.11)$$

p-Values When B_∞ Is Not a Priori Proportional to $\overline{U}_\infty$

When B_∞ is not *a priori* proportional to $\overline{U}_\infty$, two things happen. First, the analysis is no longer neat. Second, any Bayesian inference with large k and small m is going to be sensitive to the prior distribution placed on the k eigenvalues of $B_\infty \overline{U}_\infty^{-1}$ (equivalently, the k fractions of missing information, the eigenvalues of $B_\infty T_\infty^{-1}$). One approach is to still use the result appropriate when $B_\infty \propto \overline{U}_\infty$, that is, $\tilde{D}_m$ in (3.4.10) and (3.4.11). An obvious disadvantage of this answer is that as $m \to \infty$, the correct Bayesian p-value is not obtained when B_∞ is not proportional to $\overline{U}_\infty$ because (3.4.10) does not equal (3.4.3) for large m.

An *ad hoc* answer that yields the correct answer as $m \to \infty$ is supplied by simply replacing B_∞ in (3.4.3) with B_m, and then adjusting for the finiteness of m by referring the result to an $F_{k,\nu}$ reference distribution where ν is found from (3.3.11) using the average between/within ratio, r_m, given by (3.4.8). Thus, the approximation is

$$p\text{-value}(Q_0|S_m) \doteq \text{Prob}\{F_{k,\nu} > D_m\}, \qquad (3.4.12)$$

where

$$D_m = (Q_0 - \overline{Q}_m)[\overline{U}_m + (1 + m^{-1})B_m]^{-1}(Q_0 - \overline{Q}_m)^t/k. \qquad (3.4.13)$$

Obvious disadvantages of this answer are that for large k and small m, B_m is a very noisy (e.g., deficient rank) estimate of B_∞, and the reference distribution is not well motivated.

In practice, it is likely that either (a) m will be small (e.g., 5 or less) since the auxiliary data set created by the imputer giving the multiple imputations must be modest in size, or (b) m can be chosen to be very large (e.g., greater than 50 times the largest k) since the imputations are being created by the data analyst, with the consequence that B_∞ can be treated as essentially equal to B_m. Hence, although some technical work is needed on this problem, perhaps it should focus on finding accurate p-values using the

SIGNIFICANCE LEVELS FROM REPEATED SIGNIFICANCE LEVELS   99

statistics D_m, $\tilde{D}_m$, and one scalar summary of evidence about the lack of proportionality; this approach requires $m \geq 3$. A simple temporary answer is to refer $\tilde{D}_m$ to an F distribution on k and $\nu(k+1)/2$ degrees of freedom rather than k and $k\nu$ degrees of freedom as would be appropriate when it is known that B_∞ is proportional to $\bar{U}_\infty$:

$$p\text{-value}(Q_0|S_m) \doteq \text{Prob}\{F_{k,(k+1)\nu/2} > \tilde{D}_m\}; \qquad (3.4.14)$$

$(k+1)\nu/2$ is chosen simply because it is halfway between the minimum and maximum denominator degrees of freedom. Current work by Raghunathan (1987) provides suggestions for improvements. Also, a variety of other approaches are suggested by Li (1985).

3.5. SIGNIFICANCE LEVELS FROM REPEATED COMPLETED-DATA SIGNIFICANCE LEVELS

Another perspective, however, is that this problem of obtaining p-values for k-component Q from m completed-data moments when m is small is not the crucial one in practice. In practice, with multicomponent Q and a null value Q_0, complete-data analyses usually do not produce an estimate and variance–covariance matrix, but rather simply the completed-data test statistic and corresponding p-value. Thus, the repeated completed-data analyses often will not be summarized by $S_m = \{\hat{Q}_{*l}, U_{*l}, l = 1, \ldots, m\}$ but rather simply by the repeated χ^2 statistics $\{d_{*l}, l = 1, \ldots, m\}$ where asymptotically the d_{*l} are given by

$$d_{*l} = (Q_0 - \hat{Q}_{*l})U_{*l}^{-1}(Q_0 - \hat{Q}_{*l})^t \qquad (3.5.1)$$

and

$$p\text{-value in } l\text{th completed data set} = \text{Prob}\{\chi_k^2 > d_{*l}\}. \qquad (3.5.2)$$

Consequently, finding the p-value for Q_0 given the d_{*l} rather than S_m is arguably a more important practical problem.

A New Test Statistic

If we accept the propriety of $\tilde{D}_m$ as a test statistic based on S_m, it is relevant to consider the test statistic

$$\hat{D}_m = \frac{\dfrac{\bar{d}_m}{k} - \dfrac{m-1}{m+1}r_m}{1 + r_m} \qquad (3.5.3)$$

where

$$\bar{d}_m = \sum_{l=1}^{m} d_{*l/m}. \qquad (3.5.4)$$

Under the asymptotic sampling distribution with all U_{*l} essentially equal, $\hat{D}_m$ is identical to $\tilde{D}_m$ even though it depends on the k vectors of means Q_{*l} and the $k \times k$ matrices of covariances U_{*l} only through the scalars $d_{*1}, \ldots, d_{*m}$, and r_m. Of course, the dependence of $\hat{D}_m$ on r_m remains an issue.

The Asymptotic Equivalence of $\tilde{D}_m$ and $\hat{D}_m$—Proof

In order to avoid cluttering expressions, without loss of generality for this proof, suppose Q_0 is all zero and all U_{*l} equal the $k \times k$ identity matrix. Then $\tilde{D}_m$ in (3.4.11) is given by

$$(1 + r_m)^{-1}(\overline{Q}_m \overline{Q}_m^t / k) \qquad (3.5.5)$$

and $\bar{d}_m$ in $\hat{D}_m$ is given by

$$\sum_{l=1}^{m} \hat{Q}_{*l} \hat{Q}_{*l}^t / m,$$

or equivalently,

$$\overline{Q}_m \overline{Q}_m^t + \sum_{l=1}^{m} (\hat{Q}_{*l} - \overline{Q}_m)(\hat{Q}_{*l} - \overline{Q}_m)^t / m,$$

which, since $\operatorname{Tr}(AB) = \operatorname{Tr}(BA)$, gives

$$\overline{Q}_m \overline{Q}_m^t + (m - 1)\operatorname{Tr}(B_m)/m,$$

or by (3.4.8),

$$\bar{d}_m = \overline{Q}_m \overline{Q}_m^t + (m - 1)kr_m/(m + 1). \qquad (3.5.6)$$

Substituting (3.5.6) for $\bar{d}_m$ in (3.5.3) yields (3.5.5) and thus $\hat{D}_m$ is asymptotically equivalent to $\tilde{D}_m$.

Integrating over r_m to Obtain a Significance Level from Repeated Completed-Data Significance Levels

Suppose we accept the propriety of using $\tilde{D}_m$ and an associated $F_{k, k'\nu}$ reference distribution [e.g., $k' = k$ in (3.4.10) and $k' = (k + 1)/2$ in

(3.4.14)]. Then the asymptotic equivalence of $\tilde{D}_m$ and $\hat{D}_m$ implies that

$$p\text{-value}(Q_0|d_{*1},\ldots,d_{*m}) \doteq \int \Pr(F_{k,k'\nu} > \hat{D}_m)\Pr(r_m|d_{*1},\ldots,d_{*m}). \tag{3.5.7}$$

Unfortunately, but not surprisingly, the distribution of r_m given $d_{*1},\ldots,d_{*m}$ does not seem easy to work with. Ideally, a simple approximation could be found that would incorporate available prior information such as supplied by the data collector about the range of fractions of missing information or the data analyst about the fractions of information missing found when doing interval estimation for some typical components of Q.

An easy but not very satisfying approach is to replace $\Pr(r_m|d_{*1},\ldots,d_{*m})$ with point mass at some estimate of r_m and reduce the denominator degrees of freedom to reflect the number of individual estimates combined to estimate r_m; since d_{*l} is a scalar rather than a k-component quantity, this suggests reducing the degrees of freedom by the factor k^{-1}. For example, assuming $B_\infty = r_\infty \overline{U}_\infty$, a method of moments estimate of r_m suggested in Li (1985) is

$$\hat{r}_m = \frac{(1 + m^{-1})s_d^2}{2\bar{d}_m + \left[4\bar{d}_m^2 - 2ks_d^2\right]_+^{1/2}} \tag{3.5.8}$$

where

$$s_d^2 = \sum_{l=1}^{m}(d_{*l} - \bar{d}_m)^2/(m-1),$$

and

$$[a]_+ = a \text{ if } a > 0 \text{ and } 0 \text{ otherwise}.$$

The resultant test statistic is

$$\hat{D}_m = \frac{\dfrac{\bar{d}_m}{k} - \dfrac{m-1}{m+1}\hat{r}_m}{1 + \hat{r}_m}, \tag{3.5.9}$$

which is referred to an F distribution on k and $(1 + k^{-1})\hat{\nu}/2$ degrees of freedom, where

$$\hat{\nu} = (m-1)(1 + \hat{r}_m^{-1})^2, \tag{3.5.10}$$

with $\hat{r}_m$ given by (3.5.8).

More work is needed on this problem, both theoretically and in the context of real problems. Li (1985) provides many suggestions and alternatives, but much remains to be done. Work in progress by Raghunathan (1987) develops several improvements.

3.6. RELATING THE COMPLETED-DATA AND COMPLETE-DATA POSTERIOR DISTRIBUTIONS WHEN THE SAMPLING MECHANISM IS IGNORABLE

The completed-data posterior distribution for Q, which appears in all the results of this chapter as the distribution to be calculated on each data set completed by imputation, is

$$\Pr(Q|X, Y_{inc}, R_{inc}, I). \tag{3.6.1}$$

When the sampling mechanism is ignorable, as it will be in standard scientific surveys, (3.6.1) can be written as

$$\Pr(Q|X, Y_{inc}, R_{inc}). \tag{3.6.2}$$

It is easy to see that expressions (3.6.2) and (3.6.1) are equal when the sampling mechanism is ignorable. By Bayes's theorem

$$\Pr(Q|X, Y_{inc}, R_{inc}, I) = \Pr(Q|X, Y_{inc}, R_{inc}) \frac{\Pr(I|X, Y_{inc}, R_{inc}, Q)}{\Pr(I|X, Y_{inc}, R_{inc})},$$

and the ratio on the right-hand side is 1 since $Q = Q(X, Y)$ implies $\Pr(I|X, Y_{inc}, R_{inc}, Q) = \Pr(I|X, Y, R_{inc})$, and the ignorability of the sampling mechanism means that $\Pr(I|X, Y, R_{inc}) = \Pr(I|X, Y_{inc}, R_{inc}) = \Pr(I|X, Y_{obs}, R_{inc})$.

The completed-data posterior distribution given by (3.6.2) is not quite the posterior distribution that would naturally be calculated in the absence of nonresponse in a scientific survey, and thus is not quite the one data analysts would tend to form on each data set completed by imputation. In a standard scientific survey, the posterior distribution for Q that data analysts will tend to use on each data set completed by imputation ignores the distinction between real and imputed values as if there were no possibility of nonresponse, thus ignoring the value of R_{inc}. The conditional distribution of Q given (X, Y_{inc}) will be called the *complete-data posterior distribution for Q*:

$$\Pr(Q|X, Y_{inc}). \tag{3.6.3}$$

The complete-data posterior distribution (3.6.3) differs from the completed-data posterior distribution (3.6.2) in that the former does not explicitly condition on R_{inc}. Specific examples of complete-data posterior distributions for $Q = \bar{Y}$ were given in Sections 2.5 and here in Chapter 3. For instance, with a simple random sample of scalar Y_i, the standard complete-data posterior distribution for $\bar{Y}$ is normal with mean $\bar{y}$ and variance $s^2(n^{-1} - N^{-1})$.

The question to be addressed in this section is whether the extra conditioning on R_{inc} in the completed-data posterior distribution, called for in the results of Sections 3.2–3.5 makes any practical difference in standard scientific surveys. If so, users of multiply-imputed data sets would be advised to include adjustments for the differing status of real and imputed values in their analyses of each completed data set. If not, users of multiply-imputed data sets can use complete-data methods of analysis for each completed data set, as implied by the summary of Section 3.1. The following discussion supports the conclusion that using complete-data methods is appropriate. Problems 27–29 and the frequency evaluations in Chapter 4 also support this conclusion.

Result 3.3. The Completed-Data and Complete-Data Posterior Distributions Are Equal When Sampling and Response Mechanisms Are Ignorable

From Bayes's theorem we can write

$$\Pr(Q|X, Y_{inc}, R_{inc}) = \Pr(Q|X, Y_{inc}) \frac{\Pr(R_{inc}|X, Y_{inc}, Q)}{\Pr(R_{inc}|X, Y_{inc})} \quad (3.6.4)$$

completed-data posterior distribution for Q, (3.6.2) = complete-data posterior distribution for Q, (3.6.3) $\times$ adjustment factor

When nonresponse is ignorable, the adjustment factor is 1, and thus the completed-data and complete-data posterior distributions are equal.

More generally, when $\Pr(R_{inc}|X, Y_{inc}, Q)$ is relatively constant in Q, at least for values of Q such that $\Pr(Q|X, Y_{inc})$ is relatively large, the completed-data posterior distribution for Q will be nearly the same as the complete-data posterior distribution for Q. Furthermore, when the dimension of Q is relatively small, the adjustment factor should be relatively closer to 1. Specific results can be obtained for the standard case of i.i.d. models, and these show that it is generally safe to assume that the completed-data and complete-data posterior distributions are essentially equal.

Using i.i.d. Modeling

By appealing to de Finetti's theorem (Section 2.4), we can write the joint distribution of (X, Y, R) in i.i.d. form as

$$\Pr(X, Y, R) = \int \prod_{i=1}^{N} f(X_i, Y_i, R_i|\theta)\Pr(\theta)\, d\theta \qquad (3.6.5)$$

or

$$\Pr(X, Y, R) = \int \prod_{i=1}^{N} f_{XY}(X_i, Y_i|\theta_{XY}) f_{R|XY}(R_i|X_i, Y_i, \theta_{R|XY})\Pr(\theta)\, d\theta$$

$$(3.6.6)$$

where θ_{XY} and $\theta_{R|XY}$ are functions of θ; θ_{XY} is the parameter of the distribution of (X_i, Y_i) and $\theta_{R|XY}$ is the parameter of the distribution of R_i given (X_i, Y_i). Examples of such specifications for $f_{XY}(\cdot|\cdot)$ were given in Section 2.6. An example of such a specification for $f_{R|XY}(\cdot|\cdot)$ with scalar Y_i is

$$\Pr(R|Y, \theta_{R|XY}) = \prod_{i=1}^{N} \left[Y_i^2 / \left(Y_i^2 + \theta_{R|XY}^2\right)\right]^{R_i} \left[\theta_{R|XY}^2 / \left(Y_i^2 + \theta_{R|XY}^2\right)\right]^{1-R_i}.$$

Result 3.4. The Equality of Completed-Data and Complete-Data Posterior Distributions When Using i.i.d. Models

Suppose that:

1. the sampling mechanism is ignorable, and the joint distribution of (X, Y, R) is modeled in i.i.d. form (3.6.6), where

2. conditional on θ_{XY}, the completed-data and complete-data posterior distributions of Q are equal,

$$\Pr(Q|X, Y_{inc}, R_{inc}, \theta_{XY}) = \Pr(Q|X, Y_{inc}, \theta_{XY}), \qquad (3.6.7)$$

and

3. the completed-data and complete-data posterior distributions of θ_{XY} are equal,

$$\Pr(\theta_{XY}|X, Y_{inc}, R_{inc}) = \Pr(\theta_{XY}|X, Y_{inc}). \qquad (3.6.8)$$

Then the completed-data and complete-data posterior distributions of Q are equal,

$$\Pr(Q|X, Y_{inc}, R_{inc}) = \Pr(Q|X, Y_{inc}).$$

The proof of this result is immediate upon integrating the product of (3.6.7) and (3.6.8) over θ_{XY}. Its importance lies in the fact that in many cases conditions (3.6.7) and (3.6.8) can be easily seen to be true or approximately true.

Example 3.5. A Situation in Which Conditional on θ_{XY}, the Completed-Data and Complete-Data Posterior Distributions of Q Are Equal—Condition (3.6.7)

Suppose that the sampling mechanism is such that all components of Y_i are either included in or excluded from the sample, so that $I_i = (1, \ldots, 1)$ or $(0, \ldots, 0)$; this is a very common situation. Then the completed-data posterior distribution of Q given θ_{XY} is determined by (a) the observed values (X, Y_{inc}) and (b) the completed-data posterior distribution of Y_{exc} given θ_{XY}, which can be expressed as

$$\Pr(Y_{exc}|X, Y_{inc}, R_{inc}, \theta_{XY}) = \prod_{i \in ex} \frac{f_{XY}(X_i, Y_i|\theta_{XY})}{\int f_{XY}(X_i, Y_i|\theta_{XY}) \, dY_i} \quad (3.6.9)$$

where $ex = \{i | I_{ij} = 0 \text{ for all } j\}$. The right-hand side of (3.6.9) is $\Pr(Y_{exc}|X, Y_{inc}, \theta_{XY})$, whence condition (3.6.7) holds in the very common case when units are either included in or excluded from the survey.

Example 3.6. Cases in Which Condition (3.6.7) Nearly Holds

A case similar to that of Example 3.5 occurs when the population is very large relative to the sample in the sense that $Q(X, Y)$ is essentially a function of $\{(X_i, Y_i), i \in ex\}$. Then the completed-data posterior distribution of Q given θ_{XY} is determined by (3.6.9), whence condition (3.6.7) follows. A slight generalization is to have a large population with a large group of units with $I_i = (1, \ldots, 1)$ but a small group of units with $I_i \neq (1, \ldots, 1)$ or $(0, \ldots, 0)$. In this case, the completed-data posterior distribution of Q given θ_{XY} is determined by the observed values of (X, Y_{inc}) and (3.6.9), and again (3.6.7) follows. The conclusion of this example and the previous one is that condition (3.6.7) can be expected to hold quite generally in practice.

Example 3.7. Situations in Which the Completed-Data and Complete-Data Posterior Distributions of θ_{XY} Are Equal—Condition (3.6.8)

Suppose θ_{XY} and $\theta_{R|XY}$ are *a priori* independent. Then it is easy to see from (3.6.6) that (3.6.8) holds. The only reason then why (3.6.8) might not hold is *a priori* ties between θ_{XY} and $\theta_{R|XY}$.

In standard complete-data problems with large samples, standard asymptotic arguments (Section 2.10) imply that the observed data tend to overwhelm the prior distribution in the sense that essentially the same posterior distribution is obtained for a variety of prior distributions that do not place severe restrictions on parameters. Since we are assuming for each completed data set that (X, Y_{inc}, R_{inc}) is fully observed, these standard asymptotic arguments suggest that in large samples any *a priori* dependence between θ_{XY} and $\theta_{R|XY}$ will commonly make little difference to the posterior distribution of θ_{XY}.

Example 3.8. A Simple Case Illustrating the Large-Sample Equivalence of Completed-Data and Complete-Data Posterior Distributions of θ_{XY}

Suppose Y_i is binary (0–1) and there is no X_i. The i.i.d. model for (Y_i, R_i) given θ can be parametrized as

$$\theta_Y = \Pr(Y_i = 1|\theta)$$

$$\theta_{r1} = \Pr(R_i = 1|Y_i = 1, \theta)$$

$$\theta_{r0} = \Pr(R_i = 1|Y_i = 0, \theta)$$

where θ_Y is θ_{XY} and $(\theta_{r1}, \theta_{r0})$ is $\theta_{R|XY}$ in the notation of (3.6.6). If the prior distribution for $(\theta_Y, \theta_{r1}, \theta_{r0})$ has positive prior density everywhere in $(0, 1)^3$, in large samples the posterior distribution of θ_Y is normal with mean equal to the maximum-likelihood estimate of θ_Y, (i.e., $\bar{y}$), and variance equal to the negative inverse of the second derivative of the log likelihood at the maximum-likelihood estimate [i.e., $\bar{y}(1 - \bar{y})/n$], so that $\Pr(\theta_Y|Y_{inc}, R_{inc}) = \Pr(\theta_Y|Y_{inc})$. Of course, if *a priori* there are strong ties between θ_Y and $(\theta_{r1}, \theta_{r0})$, such as $\theta_Y = (\theta_{r1} + \theta_{r0})/2$, this conclusion will not hold. This example and the previous one suggest that with large samples, commonly condition (3.6.8) will be satisfied.

The General Use of Complete-Data Statistics

Examples 3.5–3.8 suggest that with large samples both conditions (3.6.7) and (3.6.8) will be approximately satisfied, and thus Result 3.4 implies that in scientific surveys, completed-data and complete-data posterior distributions will generally be practically the same. Consequently, assuming normal posterior distributions, the statistics to be calculated on each completed data set can be complete-data statistics involving the mean and variance of Q given (X, Y_{inc}) under the model $\Pr(X, Y)$ used to create the imputations.

In practice, however, the *analyst's* model for (X, Y) will often differ from the *imputer's* model for (X, Y), especially when the analyst differs from the imputer. The crucial question then becomes whether the mean and variance of Q given (X, Y_{inc}) under the analyst's model approximately equal the mean and variance of Q given (X, Y_{inc}) under the imputer's model. If so, then repeated-imputation inferences using the data analyst's complete-data statistics are approximately valid under the imputer's model $\Pr(X, Y, R, I)$. The broad applicability of the standard complete-data statistics $\hat{Q} = \bar{y}$ and $U = s^2(n^{-1} - N^{-1})$ in simple random samples for $Q = \bar{y}$ (see Examples 2.2 and 2.3) supports the contention that such approximate validity is not uncommon. Although further formal work within the Bayesian framework is desirable, our consideration of the effect of using different models for imputation and analysis will take place within the context of the randomization-based evaluations of Chapter 4.

PROBLEMS

1. Summarize the advice to the analyst of a multiply-imputed data set presented in Section 3.1.
2. When are multiple imputations repetitions?
3. In Example 3.1, calculate the p-value associated with $\hat{D}_m$ where γ_m is replaced with the average fraction of information missing for the two components of β.
4. Repeat Example 3.1 where

$$B_m = \begin{bmatrix} 20 & 10 \\ 10 & 30 \end{bmatrix} \text{ and then } B_m = \begin{bmatrix} 2 & 1 \\ 1 & 3 \end{bmatrix},$$

but the other values remain unchanged. Does the value for $\hat{D}_m$ still make sense in these cases? That is, is 15 still a plausible value for the average complete-data χ^2 statistic in both cases?
5. In Example 3.2, show directly that for infinite m, the average value of $\text{Prob}\{\bar{Y} > 0 | X, Y_{inc}\}$ equals $\text{Prob}\{\bar{Y} > 0 | X, Y_{obs}\}$.
6. In Example 3.3, suppose p-values for the null value Q_0 are calculated on each completed data set, $p_1, p_2, \ldots$. For infinite m, find the

average p-value for $Q_0 = 0$. Does this equal the correct p-value? Provide some insight regarding your answer. Hint: Distinguish between the p-value for $Q_0 = 0$ and $\text{Prob}\{Q > 0 | X, Y_{obs}, R_{inc}\}$.

7. Suppose $(Q - \hat{Q}) \sim N(0, U)$ where Q is k-dimensional, $Q = (Q_1, Q_2, \ldots, Q_k)$. Find the significance level associated with the null hypothesis $Q_1 = Q_2$. Hint: What's the distribution of $(Q_1 - Q_2 - \hat{Q}_1 + \hat{Q}_2)$ and the null value associated with the null hypothesis $Q_1 = Q_2$?

8. Define the EM algorithm (Dempster, Laird, and Rubin, 1977) and show that its rate of convergence is governed by the largest fraction of missing information.

9. Discuss the following claim: with univariate Y_i, no covariates, and a scalar estimand, the fraction of missing information essentially equals the fraction of missing values.

10. Provide rigorous justification for (3.3.2) and (3.3.3) and the consequential (3.3.7) and (3.3.8).

11. Derive (3.3.15) and the expectation and variance of its right-hand side.

12. Suppose a $100(1 - \alpha)\%$ interval estimate C is constructed using (3.3.16). From (3.3.7) and (3.3.8) show that

$$\text{Prob}\{Q \in C | X, Y_{obs}, R_{inc}, I\} = \text{Prob}\left\{z^2 < t_\nu(\alpha/2)^2 \frac{1 + r_m/q}{1 + r_m}\right\}$$

where z and q are independent $N(0,1)$ and χ^2_{m-1} random variables. Tabulate some coverages as a function of α, m, and r_m.

13. Create approximations to the distributions of Q given S_m other than (3.3.16). In particular, consider the approximation to the Behrens–Fisher distribution in Box and Tiao (1973) and the normal approximation using the Fisher (1935) factor $(\nu + 3)/(\nu + 1)$. Comment on the differing justifications for these approximations and relate these to measurements of missing information with finite m.

14. Comment on the relevance of the approach in Section 3.3 to the Bayesian analysis of general simulation experiments.

15. Show that, in general, the distribution of B_m given B_∞ is Wishart.

16. Show that when $B_\infty = r_\infty \bar{U}_\infty$ and the prior distribution on $\log(r_\infty)$ is proportional to a constant, then the distribution of r_∞ given S_m is $k(m - 1)r_m$ times an inverted χ^2 random variable.

17. Find a better reference distribution for D_m than $F_{k, \nu}$, and provide some evaluations.

18. Assume some simple structure on $(B_\infty, \bar{U}_\infty)$ other than $B_\infty \propto \bar{U}_\infty$, and derive Bayesian p-values under this assumption.

PROBLEMS

19. Show that letting Q_0 be zero and all U_{*l} equal the identity matrix results in no loss of generality when showing the asymptotic equivalence of $\tilde{D}_m$ and $\hat{D}_m$.

20. Prove that for $m \to \infty$ and $B_\infty = r_\infty \bar{U}_\infty$,

$$r_\infty = \frac{s_d^2}{2\bar{d}_\infty + [4\bar{d}_\infty - 2ks_d^2]^{1/2}}.$$

Hint: Show that

$$\bar{d}_\infty = D_\infty(1 + r_\infty) + kr_\infty$$

$$s_d^2 = 2kr_\infty^2 + 4r_\infty D_\infty(1 + r_\infty).$$

21. Prove that for $m \to \infty$ the moments of the d_{*l} can be calculated in terms of D_∞ and the k eigenvalues of B_∞ with respect to $\bar{U}_\infty$. Conclude that, in principle, with m large enough, D_∞ can be calculated as a function of $k + 1$ moments of the infinite set of repeated d_{*l} (Raghunathan, 1987).

22. Create improved methods of finding p-values from $d_{*1}, \ldots, d_{*m}$ when $B_\infty \propto \bar{U}_\infty$.

23. Extend Problem 22 to the general case.

24. Review results in Li (1985) concerning hypothesis testing beyond those presented here.

25. State precisely the difference between the completed-data posterior distribution for Q and the complete-data posterior distribution for Q. Also, summarize conditions under which this distinction can be ignored. Finally, provide a realistic example in which this distinction is important.

26. Suppose the imputer has available the values of a variable Z not available to the data analyst. Suppose that the object of inference is $Q = Q(X, Y)$, which is not a function of Z, and that the imputer's and data analyst's models are the same. Describe conditions under which the distribution of Q being simulated,

$$\int \Pr(Q|X, Y_{inc}, R_{inc})\Pr(Y_{mis}|X, Y_{obs}, R_{inc}, Z)\, dY_{mis},$$

equals or nearly equals the imputer's posterior distribution of Q:

$$\Pr(Q|X, Y_{inc}, R_{inc}, Z_{inc}).$$

Hint: First show that
$$\Pr(Y_{mis}|X, Y_{obs}, R_{inc}, Z) = \Pr(Y_{mis}|X, Y_{obs}, R_{inc}, Z_{inc}).$$
Then consider when $\Pr(Q|X, Y_{inc}, R_{inc}) \doteq \Pr(Q|X, Y_{inc}, R_{inc}, Z_{inc})$ by considering when $\Pr(Y_{exc}|X, Y_{inc}, R_{inc}) \doteq \Pr(Y_{exc}|X, Y_{inc}, R_{inc}, Z_{inc})$ and when $\Pr(\theta_{Y|X}|X, Y_{inc}, R_{inc}) \doteq \Pr(\theta_{Y|X}|X, Y_{inc}, R_{inc}, Z_{inc})$, for example, $\theta_{Y|X, R}$ *a priori* independent of $\theta_{Z|X, Y, R}$ and large samples.

27. Suppose $C = C(X, Y_{inc}, I)$ is an interval estimate of Q that is free of R and that the sampling mechanism is unconfounded with R. Show that the Bayesian calibration rate of C, given (X, Y_{inc}, I) but averaging over R_{inc}, is the same for all response mechanisms. That is, show that for fixed $\Pr(X, Y), \Pr(I|X, Y)$, and (X, Y_{inc}, I), the average posterior probability that C includes Q,

$$\text{Prob}\{Q \in C|X, Y_{inc}, I\}$$
$$= E[\text{Prob}\{Q \in C|X, Y_{inc}, R_{inc}, I\}|X, Y_{inc}, I],$$

takes the same value for all specifications $\Pr(R|X, Y)$. Hint: Write

$$\text{Prob}\{Q \in C|X, Y_{inc}, I\}$$
$$= \int\int \delta(Q \in C) \frac{\Pr(X, Y)\Pr(R|X, Y)\Pr(I|X, Y, R)\, dY_{exc}\, dR}{\int\int \Pr(X, Y)\Pr(R|X, Y)\Pr(I|X, Y, R)\, dY_{exc}\, dR}$$

and note that Q, C, and $\Pr(I|X, Y, R) = \Pr(I|X, Y)$ are free of R.

28. Consider a simple simulation with no X and scalar Y_i where the following steps are passed through 10^6 times to generate 10^6 populations of Y each of size $N = 10^3$. First, draw 10^3 $N(0, 1)$ deviates and exponentiate them to create a population where the Y_i are lognormal and $\overline{Y}$ is the population mean. Second, draw 10^5 values of R from each of two response mechanisms, the first one unconfounded,

$$\Pr(R|Y) = (0.6)^{\Sigma R_i}(0.4)^{N-\Sigma R_i},$$

and the second one confounded with Y,

$$\Pr(R|Y) = \prod_{1}^{10^3} \left[\frac{Y_i}{1-Y_i}\right]^{R_i} \left[\frac{1}{1-Y_i}\right]^{1-R_i},$$

thereby creating 2×10^5 populations of (Y, R) values all with the

PROBLEMS

same value of Y. Third, from each of the 2×10^5 populations, draw all $\binom{1000}{100}$ possible simple random samples of $n = 100$ values of (Y_i, R_i) from (Y, R), calculate the sample mean $\bar{y}$ and sample variance s^2, and then the interval estimate

$$C = \bar{y} \pm 1.96s(.01 - .001)^{1/2}$$

For our Y population, there are $2 \times 10^5 \times \binom{1000}{100}$ interval estimates, but for each of the $\binom{1000}{100}$ possible values of I, all 2×10^5 interval estimates are the same. There are 10^6 such Y populations.

Suppose for each of the $2 \times 10^5 \times 10^6 \times \binom{1000}{100}$ simple random samples we calculate the posterior probability that $\overline{Y} \in C$ under the lognormal model for Y and the correct model for the response mechanism:

$$p^u = \int\int \delta(Q \in C) \Pr^u(Y_{exc}, R_{exc} | Y_{inc}, R_{inc}, I) \, dY_{exc} \, dR_{exc},$$

and

$$p^c = \int\int \delta(Q \in C) \Pr^c(Y_{exc}, R_{exc} | Y_{inc}, R_{inc}, I) \, dY_{exc} \, dR_{exc},$$

where the superscripts u and c refer to the unconfounded and confounded response mechanisms, respectively. Now for each of the $10^6 \times \binom{1000}{100}$ simple random samples of Y, average the 10^5 values of p^u calculated for the 10^5 values of R_{inc} generated under the unconfounded response mechanism, and average the 10^5 values of p^c calculated from the 10^5 values of R_{inc} generated under the confounded response mechanism. This creates two average probabilities for each of the $10^6 \times \binom{1000}{100}$ samples of Y, which can be written, supposing 10^5 is essentially infinite, as

$$\bar{p}^u = \int p^u \Pr^u(R_{inc} | Y_{inc}, I) \, dR_{inc}$$

and

$$\bar{p}^c = \int p^c \Pr^c(R_{inc} | Y_{inc}, I) \, dR_{inc}.$$

Verify from the expressions for $\bar{p}^u$, $\bar{p}^c$, p^u, and p^c, that $\bar{p}^u = \bar{p}^c$. Reach the same conclusion, namely, that

$$\bar{p}^u = \bar{p}^c = \int \delta(Q \in C) \Pr(Y_{exc}|Y_{inc}, I) \, dY_{exc},$$

from the general result in Problem 27, and comment on the implications for using complete-data rather than completed-data statistics.

29. As in Example 3.8, suppose Y_i is binary and there is no X_i, but now let $\theta_0 = \Pr(Y_i = 1|R_i = 0, \theta)$, $\theta_1 = \Pr(Y_i = 1|R_i = 1, \theta)$, and $\theta_R = \Pr(R_i = 1|\theta)$, where $\theta = (\theta_0, \theta_1, \theta_R)$ with θ_0, θ_1, and θ_R *a priori* independent.
   (a) Show that generally θ_{XY} and $\theta_{R|XY}$ are *a priori* dependent.
   (b) Show that generally nonresponse is nonignorable.
   (c) Assume R is known in the population with N_1 respondents and N_0 nonrespondents, and show that given R the population respondent mean $\bar{Y}_1$ and the nonrespondent mean $\bar{Y}_0$ are *a posteriori* independent.
   (d) Apply the standard inference to $\bar{Y}_0$ and $\bar{Y}_1$ and deduce that given R the posterior mean of $\bar{Y}$ is nearly

$$(N_1/N)\bar{y}_1 + (N_0/N)\bar{y}_0$$

and the posterior variance of $\bar{Y}$ is nearly

$$(N_1/N)^2 s_1^2 (n_1^{-1} - N_1^{-1}) + (N_0/N)^2 s_0^2 (n_0^{-1} - N_0^{-1}),$$

in an obvious notation.
   (e) Apply the standard inference to estimate $\bar{R} = N_1/N$ and integrate the answer in (d) over this to obtain a final inference for $\bar{Y}$.
   (f) Comment on the relationship between this problem and Example 3.8.

30. Extend Problem 29 to the case where under the imputer's model the Y_i have a normal linear regression on X_i with slopes depending on R_i. Consider using the following models to form a repeated imputation inference for $\bar{Y}$ based on complete-data statistics: the imputer's model, a normal linear regression model, and a quadratic regression model.

CHAPTER 4

Randomization-Based Evaluations

4.1. INTRODUCTION

The repeated imputation inferences developed in Chapter 3 are approximations derived in the idealized situation where the imputer and analyst use a common Bayesian model. To ensure that these inferences are approximately calibrated in a wide variety of routine applications, it is appropriate and in the tradition of survey methodology to evaluate their performance from the randomization perspective. We restrict attention to standard scientific surveys and standard complete-data statistics. Consequently, our random-response randomization-based evaluations treat X and Y as fixed, assume specifications for the response mechanism $\Pr(R|X, Y)$ and the unconfounded sampling mechanism $\Pr(I|X)$, and calculate the frequency properties of repeated-imputation inferences using complete-data statistics averaging over I and R for fixed (X, Y).

Major Conclusions

Our major conclusions regarding these evaluations are as follows. First, if (1) the complete-data inference is a valid randomization-based inference in the absence of nonresponse and (2) the imputation method is proper, as defined in Section 4.2, then in large samples the repeated-imputation procedures lead to inferences that are valid from the random-response randomization-based perspective, at least when the number of imputations is large. Furthermore, under these same two conditions, when the number of imputations, m, is small, the procedures are nearly valid from the random-response randomization-based perspective, and their operating characteristics can be relatively easily calculated.

TABLE 4.1. Large-sample relative efficiency (in %)[a] when using a finite number of proper imputations, m, rather than an infinite number, as a function of the fraction of missing information, γ_0: RE = $(1 + \gamma_0/m)^{-1/2}$.

	γ_0								
m	.1	.2	.3	.4	.5	.6	.7	.8	.9
1	95	91	88	85	82	79	77	75	73
2	98	95	93	91	89	88	86	85	83
3	98	97	95	94	93	91	90	89	88
5	99	98	97	96	95	94	94	93	92
∞	100	100	100	100	100	100	100	100	100

[a] in units of standard deviations

Large-Sample Relative Efficiency of Point Estimates

As an example, the large-sample variance of the repeated-imputation point estimator $\overline{Q}_m$ based on a proper imputation method is

$$V(\overline{Q}_m | X, Y) = (1 + \gamma_0/m) V(\overline{Q}_\infty | X, Y),$$

where γ_0 is the population fraction of missing information. This result is derived in Section 4.3. The quantity γ_0 is equal to the expected fraction of observations missing in the simple case of scalar Y_i with no covariates, and commonly is less than the fraction of observations missing when there are covariates that predict Y_i. Thus, the efficiency of the finite-m repeated-imputation estimator relative to the fully efficient infinite-m repeated-imputation estimator is $(1 + \gamma_0/m)^{-1/2}$ in units of standard errors; values are given in Table 4.1. In cases with little missing information, proper imputation with $m = 2$ or 3 is nearly fully efficient.

Large-Sample Coverage of t-Based Interval Estimates

Similarly, it is relatively straightforward to derive the random-response randomization-based coverages of repeated-imputation interval estimates assuming valid complete-data inferences and proper imputation methods. Section 4.3 derives the large-sample results given in Table 4.2, which includes results for single imputation based on the normal reference distribution to discourage the use of single imputation. Table 4.2 suggests that proper multiple imputation with $m = 2$ or 3 will yield accurate coverages in many cases.

INTRODUCTION

TABLE 4.2. Large-sample coverage probability (in %) of interval estimates based on the t reference distribution, (3.1.8), as a function of the number of proper imputations, $m \geq 2$; the fraction of missing information, γ_0; and the nominal level, $1 - \alpha$. Also included for contrast are results based on single imputation, $m = 1$, using the complete-data normal reference distribution (3.1.1) with $\hat{Q}$ replaced by $\overline{Q}_1 = \hat{Q}_{*1}$ and U replaced by $\overline{U}_1 = U_{*1}$.

						γ_0				
$1-\alpha$	m	.1	.2	.3	.4	.5	.6	.7	.8	.9
50%	1	46	42	38	34	30	26	22	18	12
	2	50	50	51	51	50	50	50	50	50
	3	50	50	50	50	50	50	50	50	50
	5	50	50	50	50	50	50	50	50	50
	∞	50	50	50	50	50	50	50	50	50
80%	1	75	70	65	60	54	48	41	33	23
	2	80	80	80	79	78	77	76	76	76
	3	80	80	80	80	79	79	79	79	79
	5	80	80	80	80	80	80	80	80	80
	∞	80	80	80	80	80	80	80	80	80
90%	1	86	82	77	72	66	59	51	42	29
	2	90	90	89	88	87	86	85	84	83
	3	90	90	90	89	89	88	88	88	88
	5	90	90	90	90	90	90	89	89	89
	∞	90	90	90	90	90	90	90	90	90
95%	1	92	89	85	80	74	67	59	49	35
	2	95	95	94	93	92	91	89	88	87
	3	95	95	95	94	94	93	93	92	92
	5	95	95	95	95	95	94	94	94	94
	∞	95	95	95	95	95	95	95	95	95
99%	1	98	96	94	91	86	80	72	61	45
	2	99	99	98	98	97	96	95	93	92
	3	99	99	99	98	98	98	97	97	96
	5	99	99	99	99	99	99	98	98	98
	∞	99	99	99	99	99	99	99	99	99

Outline of Chapter

Because the multiple-imputation procedures we are considering are derived from the Bayesian perspective but are being evaluated from the random-response randomization-based perspective, the arguments that are used in this chapter are relatively demanding in that they rely on a clear conceptual understanding of both perspectives. In order to simplify the presentation, conditions for the validity of infinite-m multiple-imputation procedures from the randomization perspective are considered before any evaluations of finite-m multiple-imputation procedures. As a result, proper imputation methods, which essentially are multiple-imputation procedures such that the statistics $(\overline{Q}_\infty, \overline{U}_\infty, B_\infty)$ provide valid random-response randomization-based inferences for the complete-data statistics $(\hat{Q}, U)$, receive extended discussion and illustration. Following this discussion, the asymptotic distribution of $(\overline{Q}_m, \overline{U}_m, B_m)$ is derived assuming proper imputation methods. Then finite-m proper multiple-imputation procedures are evaluated, first for scalar Q and then for multicomponent Q. Related work within different conceptual structures appears in Li (1985) and Schenker and Welsh (1986), the latter being the most mathematically rigorous.

4.2. GENERAL CONDITIONS FOR THE RANDOMIZATION-VALIDITY OF INFINITE-m REPEATED-IMPUTATION INFERENCES

Suppose inferences are drawn from an infinite number of multiple imputations under one model using standard complete-data statistics, $\hat{Q} = \hat{Q}(X, Y_{inc}, I)$ and $U = U(X, Y_{inc}, I)$, and the normal reference distribution (3.2.13). The resulting repeated-imputation inferences will be valid from the random-response randomization-based perspective under the posited response mechanism and specified sampling mechanism if

$$(\overline{Q}_\infty | X, Y) \sim N(Q, T_0) \qquad (4.2.1)$$

and

$$(T_\infty | X, Y) \sim (T_0, \ll T_0), \qquad (4.2.2)$$

where $Q = Q(X, Y)$ and $T_0 = T_0(X, Y)$ are fixed by the true values of (X, Y); $T_\infty = \overline{U}_\infty + B_\infty$; and $\overline{Q}_\infty, \overline{U}_\infty$, and B_∞ are functions of (X, Y_{obs}, R_{inc}, I). The random variable in (4.2.1) and (4.2.2) is (R, I), and the $\ll$ notation was defined in Section 2.10. When (4.2.1) and (4.2.2) hold, we will say that the repeated-imputation inference is *randomization-valid*. For instance, the 95% interval estimate given by (3.2.14) will be a 95% confidence interval, and if $Q = Q_0$, then the p-value given by (3.2.15) will be uniformly distributed on $(0, 1)$.

It is important to realize that all that is required for the inference to be randomization-valid is that (4.2.1) and (4.2.2) hold under the *posited* response mechanism, *specified* sampling mechanism, and the *true* population data (X, Y). The central theme of multiple imputation is to allow the display sensitivity to different possible models for nonresponse by deriving valid inferences under each posited model for nonresponse. Consequently, there is no need to have inferences that are valid under the true response mechanism, just under the posited one.

Conditions (4.2.1) and (4.2.2) will hold in large samples if (a) repeated imputations are drawn under the true model for the response mechanism and the true model for the data and (b) the complete-data inference equals the completed-data inference derived under the same models. This conclusion follows from the general arguments in Section 2.10 regarding the asymptotic randomization-based validity of Bayesian procedures derived under the model that generated the data. Results in Section 2.10 also suggest that this conclusion can still hold when an incorrect Bayesian model is used to create both the complete-data inference and the repeated imputations; specifically, if $\overline{Q}_\infty$ is asymptotically unbiased for Q, $E(\overline{Q}_\infty | X, Y) = Q$, then conditions (4.2.1) and (4.2.2) will be satisfied asymptotically under mild regularity conditions.

Complications in Practice

Although these results are encouraging, there exist complications in practice. For example, even assuming explicit Bayesian models will be used to create repeated imputations, generally the analyst's complete-data statistics $\hat{Q}$ and U will not be the conditional mean and variance of Q given (X, Y_{inc}, R_{inc}) under the imputer's model for the data; in fact, this is nearly certain when the imputer and the data analyst are different individuals or organizations. Also, in practice the multiple imputations may not be repetitions under any particular explicit Bayesian model but rather may be approximate repetitions under an implicit model, such as a modification of the hot-deck. This will often be true in current practice when multiple imputations are created by an organization using a simple modification of an existing single-imputation procedure.

More General Conditions for Randomization-Validity

Thus, it is practically important to consider more general circumstances under which conditions (4.2.1) and (4.2.2) are satisfied. Two conditions are sufficient for the infinite-m repeated-imputation inference to be randomization-valid.

First, the complete-data inference has to be randomization-valid:

$$(\hat{Q}|X, Y) \sim N(Q_0, U_0) \quad (4.2.3)$$

and

$$(U|X, Y) \sim (U_0, \ll U_0), \quad (4.2.4)$$

where $Q_0 = Q(X, Y)$ and $U_0 = U_0(X, Y)$ are fixed by the true values of X and Y, and the underlying random variable in (4.2.3) and (4.2.4) is I with distribution $\Pr(I|X, Y) = \Pr(I|X)$. Because our concern is with problems of nonresponse, we always regard the complete-data inference to be randomization-valid.

The second condition is that the multiple-imputation procedure is *proper*, basically in the sense that the infinite-m statistics $(\overline{Q}_\infty, \overline{U}_\infty, B_\infty)$ yield randomization-valid inferences for the complete-data statistics $\hat{Q}$ and U under the posited response mechanism. For notational simplicity in the following definition, we assume that nonresponse affects all components of $\hat{Q}$ (i.e., B_∞ is full rank); the results can be seen to hold otherwise by decomposing $\hat{Q}$ into components affected by nonresponse (corresponding submatrix of B_∞ is full rank) and unaffected by nonresponse (corresponding submatrix of B_∞ is all zero).

Definition: Proper Multiple-Imputation Methods

A multiple-imputation procedure is proper for the set of complete-data statistics $\{\hat{Q}, U\}$ if three conditions are satisfied:

1. Treating (X, Y, I) as fixed, under the posited response mechanism, the $m = \infty$ multiple-imputation procedure provides randomization-valid inferences for the complete-data statistic $\hat{Q} = \hat{Q}(X, Y_{inc}, I)$ based on the statistics $\overline{Q}_\infty$ and B_∞:

$$(\overline{Q}_\infty|X, Y, I) \sim N(\hat{Q}, B) \quad (4.2.5)$$

$$(B_\infty|X, Y, I) \sim (B, \ll B), \quad (4.2.6)$$

where $B = B(X, Y_{inc}, I)$ is defined by

$$B = V(\overline{Q}_\infty|X, Y, I). \quad (4.2.7)$$

2. Treating (X, Y, I) as fixed, under the posited response mechanism, the $m = \infty$ imputation estimate of the complete-data statistic $U =$

INFINITE-m REPEATED IMPUTATION INFERENCES

$U(X, Y_{inc}, I)$, that is, $\overline{U}_\infty$, is centered at U with variability of a lower order than that of $\overline{Q}_\infty$:

$$(\overline{U}_\infty | X, Y, I) \sim (U, \ll B). \tag{4.2.8}$$

3. Treating (X, Y) as fixed, over repeated samples the variability of B is of lower order than that of $\hat{Q}$:

$$(B | X, Y) \sim (B_0, \ll U_0), \tag{4.2.9}$$

where $B_0 = B_0(X, Y)$ is defined by

$$B_0 = E(B | X, Y), \tag{4.2.10}$$

and $U_0 = U_0(X, Y)$ is defined by (4.2.4).

The underlying random variable in (4.2.5)–(4.2.8) is R with distribution specified by the response mechanism $\Pr(R | X, Y)$, whereas the underlying random variable in (4.2.9) and (4.2.10) is I with distribution $\Pr(I | X)$.

The concept of proper imputation methods deserves extended discussion and illustration, but before giving this in Section 4.3, we prove that it, when coupled with the randomization-validity of the complete-data inference, implies the randomization-validity of the infinite-m repeated-imputation inference, so that the importance of proper imputation methods is clear.

Result 4.1. If the Complete-Data Inference Is Randomization-Valid and the Multiple-Imputation Procedure Is Proper, Then the Infinite-m Repeated-Imputation Inference Is Randomization-Valid under the Posited Response Mechanism

The claim is that (4.2.3)–(4.2.10) imply (4.2.1) and (4.2.2). To prove this result, first note that (4.2.3) and (4.2.5) imply

$$(\overline{Q}_\infty | X, Y) \sim N(Q, U_0 + E(B | X, Y)),$$

which by (4.2.9) gives

$$(\overline{Q}_\infty | X, Y) \sim N(Q, U_0 + B_0).$$

Next note that by (4.2.6) and (4.2.8)

$$(\overline{U}_\infty + B_\infty | X, Y, I) \sim (U + B, \ll 2B),$$

which by (4.2.4) and (4.2.9) gives

$$(\overline{U}_\infty + B_\infty | X, Y) \sim (U_0 + B_0, \ll (2B_0 + 2U_0)),$$

as required to complete the proof.

4.3. EXAMPLES OF PROPER AND IMPROPER IMPUTATION METHODS IN A SIMPLE CASE WITH IGNORABLE NONRESPONSE

Consider the following simple situation. A simple random sample of size n is drawn from a population of N values with finite mean $Q = \overline{Y}$ and variance S^2. In the absence of nonresponse, inference for $\overline{Y}$ is based on the standard complete-data statistics $\hat{Q} = \bar{y}$, the sample mean, and $U = s^2(n^{-1} - N^{-1})$, where s^2 is the sample variance. We assume that N and n are large enough to make standard complete-data inferences based on $\hat{Q}$ and U randomization-valid, so that we accept (4.2.3) and (4.2.4) with $U_0 = S^2(n^{-1} - N^{-1})$.

Due to unconfounded nonresponse, only n_1 of the n values in Y_{inc} are observed where n_1 is also large and for simplicity is treated as fixed. The observed sample mean is $\bar{y}_1$ and sample variance is s_1^2. Since n_1 and n are large, treating the sample as fixed and averaging over the response mechanism, which is like a level of simple random sampling from the n values comprising Y_{inc}, gives for (4.2.3) and (4.2.4)

$$(\bar{y}_1 | X, Y, I) \sim N\left(\bar{y}, s^2\left(n_1^{-1} - n^{-1}\right)\right) \tag{4.3.1}$$

and

$$(s_1^2 | X, Y, I) \sim (s^2, \ll s^2). \tag{4.3.2}$$

The examples of multiple-imputation procedures that follow all concern this situation. For each example, we decide whether the imputation method is proper or improper for the complete-data statistics $\{\bar{y}, s^2(n^{-1} - N^{-1})\}$ by considering expressions (4.2.5), (4.2.6), (4.2.8), and (4.2.9).

Example 4.1. Simple Random Multiple Imputation

Consider the multiple-imputation version of the single imputation hot-deck of Example 1.6 in which multiple imputations are created by a simple random drawing with replacement from the n_1 observed values. It is easy to

EXAMPLES OF PROPER AND IMPROPER IMPUTATION METHODS 121

see that $\overline{Q}_\infty = \bar{y}_1$, so that from (4.3.1),

$$(\overline{Q}_\infty | X, Y, I) \sim N(\hat{Q}, B), \qquad (4.3.3)$$

where B, defined to be $V(\overline{Q}_\infty | X, Y, I)$, is given by

$$B = s^2(n_1^{-1} - n^{-1}). \qquad (4.3.4)$$

Thus, (4.2.5) is satisfied.

Also, it is easy to see that the average of the complete-data sample variances equals

$$s_1^2(1 - n_1^{-1})\left[1 + n_1 n^{-1}(n - 1)^{-1}\right],$$

which for large n_1 and n equals s_1^2 (see Problems 11–14 in Chapter 1 for details). Thus, for large samples,

$$\overline{U}_\infty = s_1^2(n^{-1} - N^{-1}), \qquad (4.3.5)$$

and from (4.3.2) and (4.3.4) it follows that

$$(\overline{U}_\infty | X, Y, I) \sim (U, \ll B) \qquad (4.3.6)$$

where

$$U = s^2(n^{-1} - N^{-1}), \qquad (4.3.7)$$

so that (4.2.8) is satisfied. Also, (4.2.9) is satisfied with large n, since from (4.3.4) and (4.3.7)

$$V(B|X, Y) \ll E(U|X, Y).$$

It remains to consider (4.2.7), that is, whether B_∞, the variance of the complete-data estimates $\bar{y}_{*l}$, $l = 1, 2, \ldots$ across the multiple imputations, essentially equals B, the actual variance of the complete-data estimate $\overline{Q}_\infty = \lim_{m \to \infty} \Sigma_1^m \bar{y}_{*l}/m$ over the response mechanism, given by (4.3.4). It is easy to show (see Problems 11–14 in Chapter 1 for details) that

$$B_\infty = (1 - n_1/n)(1 - n_1^{-1})s_1^2/n,$$

so from (4.3.4), for large n_1

$$E(B_\infty | X, Y, I) = Bn_1/n. \qquad (4.3.8)$$

Thus B_∞ underestimates B by the response rate, and thus (4.2.7) is not satisfied.

The conclusion is that the multiple-imputation hot-deck, which randomly samples nonrespondents from respondents, is improper for $\{\bar{y}, s^2(n^{-1} - N^{-1})\}$ for any population of Y values because it does not have enough between-imputation variability. A simple calculation therefore shows that the repeated-imputation estimated variance of $\bar{Q}_\infty$ underestimates the actual variance of $\bar{Q}_\infty$, and thus the analysis methods of Chapter 3 lead to interval estimates that are systematically too short and p-values that are systematically too significant. More precisely, for large m,

$$E(T_\infty|X, Y) = E(\bar{U}_\infty|X, Y) + E(B_\infty|X, Y), \qquad (4.3.9)$$

which from (4.3.4), (4.3.5) and (4.3.8) gives for large samples,

$$E(T_\infty|X, Y) = S^2(n^{-1} - N^{-1}) + S^2(1 - n_1/n)/n,$$

or since $V(\bar{Q}_\infty|X, Y) = S^2(n_1^{-1} - N^{-1})$,

$$E(T_\infty|X, Y) = V(\bar{Q}_\infty|X, Y) - (1 - n_1/n)S^2(n_1^{-1} - n^{-1}),$$

which shows that T_∞ underestimates $T_0 = V(\bar{Q}_\infty|X, Y)$, and so (4.2.2) fails to hold.

Why Variability Is Underestimated Using the Multiple-Imputation Hot-Deck

At first it may appear rather surprising that randomly drawing multiple imputations from respondents does not lead to valid inferences even with large samples and large m. A useful way to understand the deficiency of simple random (hot-deck) multiple imputation is to consider the problem carefully from the Bayesian point of view developed in Chapter 3.

From this perspective, the objective of multiple imputation under one model is to simulate the posterior distribution of missing Y values and thereby the posterior distribution of the statistics used for inference with complete data. This simulation is easily done in a two-step process with standard i.i.d. models: First, draw the parameters of the model from their posterior distribution, and then draw the missing values from their posterior distribution conditionally given the drawn values of the parameters.

Simple random multiple imputation omits the step of drawing the parameters and instead acts as if the respondents' distribution of Y values were exactly the same as the population distribution of Y values. Because we do not know precisely what the population values are, when we act as if

EXAMPLES OF PROPER AND IMPROPER IMPUTATION METHODS 123

we have this knowledge we underestimate variability. In other words, even if the respondents and nonrespondents represent a random partition of the population of units, $\bar{y}_1$ and s_1^2 are not the exactly appropriate mean and variance for the $n - n_1$ nonrespondents in the sample. With ignorable nonresponse, the respondents and nonrespondents share the same parameters, but the sample mean and sample variance for respondents are not perfect estimates of these parameters, and our imputations must reflect this uncertainty to be proper.

Example 4.2. Fully Normal Bayesian Repeated Imputation

Consider the fully normal repeated-imputation method of Example 3.2. It is easy to see that $\bar{Q}_\infty = \bar{y}_1$ so that (4.3.3) and (4.3.4) still hold, and thus (4.2.5) is satisfied. Also, it is easy to see that for large samples, the average of the complete-data variances equals s_1^2, so that (4.3.5)–(4.3.7) hold, and thus (4.2.8) and (4.2.9) are satisfied. Furthermore, straightforward algebra shows that in large samples

$$B_\infty = s_1^2 \left(n_1^{-1} - n^{-1} \right), \tag{4.3.10}$$

which is unbiased for B with low-order variability thus satisfying (4.2.7).

Our conclusion is that fully normal Bayesian repeated imputation is proper for $\{\bar{y}_1, s^2(n^{-1} - N^{-1})\}$ in large samples with ignorable nonresponse for any population of Y values. Substituting (4.3.5) and (4.3.10) into (4.3.9) shows that

$$E(T_\infty | X, Y) = V(\bar{Q}_\infty | X, Y)$$

to confirm the conclusion of Result 4.1 that randomization-valid inferences will result.

Example 4.3. A Nonnormal Bayesian Imputation Procedure That Is Proper for the Standard Inference—The Bayesian Bootstrap

Example 4.2 might be seen as suggesting that when the standard inference is to be used with complete data and nonresponse is ignorable, then multiple imputations should be created as repetitions under a Bayesian normal model. Many Bayesian models besides the normal, however, approximately yield the standard inference with complete data, and thus many such models can be used to create proper imputations for $\{\bar{y}, s^2(n^{-1} - N^{-1})\}$ with ignorable nonresponse.

To illustrate this, consider the Bayesian bootstrap (BB) specification of Example 2.3. A simple way to generate repeated imputations of Y_{mis} is to repeat the following two steps m independent times. For notational convenience we let $Y_{obs} = (Y_1, \ldots, Y_{n_1})$.

Step 1. Draw $n_1 - 1$ uniform random numbers between 0 and 1, and let their ordered values be $a_1, \ldots, a_{n_1-1}$; also let $a_0 = 0$ and $a_{n_1} = 1$.

Step 2. Draw each of the n_0 missing values in Y_{mis} by drawing from $Y_1, \ldots, Y_{n_1}$ with probabilities $(a_1 - a_0), (a_2 - a_1), \ldots, (1 - a_{n_1-1})$; that is, independently n_0 times, draw a uniform random number u, and impute Y_i if $a_{i-1} < u \leq a_i$.

Straightforward algebra shows that for large samples $\overline{Q}_\infty = \bar{y}_1$, $\overline{U}_\infty = s_1^2(n^{-1} - N^{-1})$, and $B_\infty = s_1^2(n_1^{-1} - n^{-1})$, and so Bayesian bootstrap imputation is asymptotically proper for $\{\bar{y}, s^2(n_1^{-1} - N^{-1})\}$ with ignorable nonresponse.

Example 4.4. An Approximately Bayesian Yet Proper Imputation Method—The Approximate Bayesian Bootstrap

Non-Bayesian imputation methods can also be proper if they incorporate appropriate between-imputation variability. Such methods are readily constructed as modifications to Bayesian methods that generate approximate repetitions. Approximate Bayesian bootstrap (ABB) imputation first draws n_1 values randomly with replacement from Y_{obs} to create Y_{obs}^*, and then draws the $n_0 = n - n_1$ components of Y_{mis} randomly with replacement from Y_{obs}^*. The difference between the ABB and the BB is that the underlying parameter of the data, which gives the probabilities of each component in Y_{obs}, is being drawn from a scaled multinomial with the ABB rather than a Dirichlet distribution. These distributions have the same means and correlations and differ in variances only by the factor $(1 + n_1^{-1})$ (see Rubin, 1981a). Consequently, repeated ABB imputation is also proper for $\{\bar{y}, s^2(n^{-1} - N^{-1})\}$ with unconfounded nonresponse in large samples.

Example 4.5. The Mean and Variance Adjusted Hot-Deck

A minor non-Bayesian modification of the fully normal imputation method of Examples 3.2 and 4.2 gives the mean and variance (MV) adjusted hot-deck. Here the first two steps of drawing μ and σ^2 from their posterior distribution are exactly as in Example 3.2. The third step in Example 3.2 is modified so that the zero-mean unit-variance z_i are no longer drawn from the normal distribution but rather with replacement from Y_{obs} linearly transformed to have zero mean and unit variance [i.e., for each component in Y_{obs}, subtract $\bar{y}_1$ and divide by $s_1(1 - n_1^{-1})^{1/2}$]. The potential advantage of this modification over fully normal imputation is that the values generated will have the same distributional shape as the observed values in Y_{obs}. For example, if the components of Y_{obs} are skew, then the MV method will tend to impute values in Y_{mis} with the same skewness, whereas the fully normal method tends to impute values in Y_{mis} that are symmetric no matter what the shape of the distribution in Y_{obs}. The methods have the same first two moments, and thus MV is asymptotically proper for $\{\bar{y}, s^2(n^{-1} - N^{-1})\}$ with unconfounded nonresponse.

A slight modification of the MV method is also proper under the same conditions: Instead of drawing σ^2 from its χ^2 posterior distribution in the first step, simply fix it at the observed variance of the components in Y_{obs}. This method was analyzed in detail by Rubin (1979a) and Herzog and Rubin (1983). Of course, if the step of drawing μ from its posterior distribution is replaced by fixing μ at the sample mean, all between-imputation variability is eliminated, and we have returned to the simple multiple-imputation hot-deck, which was shown in Example 4.1 to be improper for $\{\bar{y}, s^2(n^{-1} - N^{-1})\}$.

4.4. FURTHER DISCUSSION OF PROPER IMPUTATION METHODS

In light of Result 4.1, it obviously is desirable to use imputation methods that are proper under the posited response mechanism for as wide a variety of complete-data statistics as possible. In this section heuristic arguments and the examples of Section 4.3 are used to support the following conclusion, which suggests specific issues to be considered when choosing a multiple-imputation method in practice.

Conclusion 4.1. Approximate Repetitions from a Bayesian Model Tend to Be Proper

If imputations are drawn to approximate repetitions from a Bayesian posterior distribution of Y_{mis} under the posited response mechanism and an appropriate model for the data, then in large samples the imputation method is proper. By "an appropriate model for the data" is meant one such that $(\bar{Q}_\infty, \bar{U}_\infty)$, the posterior mean of $(\hat{Q}, U)$, is approximately unbiased for $(\hat{Q}, U)$ under the posited response mechanism:

$$E(\overline{Q}_\infty | X, Y, I) \doteq \hat{Q} \qquad (4.4.1)$$

$$E(\overline{U}_\infty | X, Y, I) \doteq U. \qquad (4.4.2)$$

There is little doubt that if this conclusion were formalized in a particular way, exceptions to it could be found. Its usefulness is not as a general mathematical result, but rather as a guide to practice. Nevertheless, in order to understand why it may be expected to hold relatively generally, it is important to provide a general heuristic argument for it.

The Heuristic Argument

We now show that drawing repetitions from appropriate Bayesian models tends to lead to proper imputation methods; that is, drawing repeated imputations from Bayesian models satisfying (4.4.1) and (4.4.2) tends to imply (4.2.5), (4.2.6), (4.2.8), and (4.2.9). Expression (4.2.8) is satisfied by the usual asymptotic argument concerning the lower-order variability of variances of estimates. Now recall that by construction B_∞ is the posterior variance of $\hat{Q}$, $V(\hat{Q}|X, Y_{obs}, R_{inc}, I)$. Results referenced in Section 2.10 suggest that if (4.4.1) holds, then (a) the distribution of $(\overline{Q}_\infty - \hat{Q})$ given (X, Y, I) will be normal [i.e., (4.2.5) will hold], and (b) B_∞ will be approximately unbiased for its variance with low-order variability [i.e., (4.2.6) will hold]. It is important to realize that if the imputation method is such that (4.4.1) and (4.4.2) are satisfied but the imputations are not drawn following the Bayesian paradigm, that is, are not repetitions, the imputation procedure can easily be improper because B_∞ may not be approximately unbiased for B as required by (4.2.6). This is illustrated by the hot-deck imputation method of Example 4.1. Expression (4.2.9) is basically a consequence of large samples since, when using reasonable imputation methods, B should be of the order of the inverse number of respondents, as in (4.3.4), and so should have lower-order variability than $\hat{Q}$ over repeated samples.

Messages of Conclusion 4.1

There are three main practical messages in Conclusion 4.1 concerning multiple-imputation procedures:

1. Draw imputations following the Bayesian paradigm as repetitions from a Bayesian posterior distribution of the missing values under the chosen models for nonresponse and data, or an approximation to this

FURTHER DISCUSSION OF PROPER IMPUTATION METHODS

posterior distribution that incorporates appropriate between-imputation variability.

2. Choose models of nonresponse appropriate for the posited response mechanism.
3. Choose models for the data that are appropriate for the complete-data statistics likely to be used—if the model for the data is correct, then the model is appropriate for all complete-data statistics.

The importance of being Bayesian or approximately so was emphasized by the examples in Section 4.3, in particular by the contrast between the hot-deck of Example 4.1 and the other imputation methods that are proper for $\{\bar{y}, s^2(n^{-1} - N^{-1})\}$ under ignorable nonresponse. All of these examples, however, assumed the complete-data statistics were $\{\bar{y}, s^2(n^{-1} - N^{-1})\}$ and assumed ignorable nonresponse, and thus did not really address the second and third messages of Conclusion 4.1.

The Importance of Drawing Repeated Imputations Appropriate for the Posited Response Mechanism

In fact, Examples 4.1–4.5 might seem to suggest that if the imputations are repetitions or approximate repetitions from a Bayesian posterior distribution, then the imputation method is always proper. But this is not true: the imputations must be appropriate for the posited response mechanism. For instance, none of the imputation methods in these examples is proper for the nonignorable response mechanism of Section 1.6 with

$$(\text{nonignorable value}) = 1.2 \times (\text{ignorable value});$$

specific examples of such mechanisms appear in Chapter 6. If nonrespondents are being considered to be systematically different from respondents, then imputations must reflect this systematic difference if the imputation method is to be proper.

The Role of the Complete-Data Statistics in Determining Whether a Repeated-Imputation Method Is Proper

The complete-data statistics play a somewhat more subtle role than the response mechanism in determining whether a repeated-imputation method is proper. Suppose in the context of Examples 4.1–4.5 with unconfounded nonresponse, the estimand had been $Q_e = \text{average} \exp(Y_i)$, where the complete-data inference for Q_e had been the standard inference applied to the n_1 observed values of $\exp(Y_i)$, which is valid in large enough samples. In this case, the complete-data statistics would be $\hat{Q}_e$ = average observed $\exp(Y_i)$, and U_e = [variance of observed $\exp(Y_i)$] $\times (n^{-1} - N^{-1})$.

It is easy to see that if the N values of Y_i were drawn from a normal distribution, then in large samples the fully normal (FN) and MV methods applied to the Y_i would be proper for $\{\hat{Q}_e, U_e\}$ or any other set of complete-data statistics. In contrast, if the Y_i had some other distribution, the FN and MV methods applied to the Y_i would generally not be proper $\{\hat{Q}_e, U_e\}$.

On the other hand, the BB and ABB methods remain proper for $\{\hat{Q}_e, U_e\}$ for any distribution in large samples because of the methods' invariance under transformations of the data. But they would not be proper for a variety of other complete-data statistics, such as the maximum observed value, except under special distributional assumptions.

This illustration should not be seen as unequivocally supporting one type of imputation method over another. In practice, large-sample arguments have to be carefully considered. Some limited evidence is provided on this point in Section 4.6.

4.5. THE ASYMPTOTIC DISTRIBUTION OF $(\overline{Q}_m, \overline{U}_m, B_m)$ FOR PROPER IMPUTATION METHODS

As proved in Section 4.2 and illustrated in Sections 4.3 and 4.4, when the multiple-imputation method is proper and the complete-data inference is randomization-valid, $m = \infty$ normal-based inferences using $(\overline{Q}_\infty, \overline{U}_\infty, B_\infty)$ are randomization-valid. Of course, in practice, an infinite number of repeated imputations under each model is impossible to provide, and so the issue of the performance of finite-m procedures immediately arises. Thus, we need to derive the joint sampling distribution of $(\overline{Q}_m, \overline{U}_m, B_m)$ over the sampling mechanism and response mechanism for fixed (X, Y).

Validity of the Asymptotic Sampling Distribution of S_m

For general evaluation purposes, we assume (i) proper imputation methods, (ii) valid complete-data inferences, and (iii) samples large enough that the asymptotic sampling distribution of S_m assumed in Chapter 3 is valid. That is, in addition to (4.2.3)–(4.2.10), we assume (3.3.2) and (3.3.3) hold, which since the $\hat{Q}_{*l}$ and U_{*l} depend on (Y, R) only through (Y_{obs}, R_{inc}), can be written as

$$(\hat{Q}_{*l} | X, Y, I, R) \sim N(\overline{Q}_\infty, B_\infty) \qquad (4.5.1)$$

$$(U_{*l} | X, Y, I, R) \sim (\overline{U}_\infty, \ll B_\infty), \qquad (4.5.2)$$

where all the $\hat{Q}_{*l}$ and U_{*l} are mutually independent given (X, Y, I, R).

ASYMPTOTIC DISTRIBUTIONS FOR PROPER IMPUTATION METHODS

When the imputations are repetitions drawn from the posterior distribution of Y_{mis} under specified models for $\Pr(X, Y)$ and $\Pr(R|X, Y)$, the arguments in Section 2.10 and 3.3 concerning the sampling distribution of posterior means ($\hat{Q}_{*l}$) and variances (U_{*l}) suggest that (4.5.1) and (4.5.2) will hold regardless of the models' correctness. Furthermore, some evidence suggests that these distributional forms will hold asymptotically for a wide variety of approximately Bayesian imputation models. Consequently, we regard (4.5.1) and (4.5.2) as generally satisfied, at least in a large-sample sense.

The Distribution of $(\overline{Q}_m, \overline{U}_m, B_m)$ Given (X, Y) for Scalar Q

The derivation of the conditional distribution of $(\overline{Q}_m, \overline{U}_m, B_m)$ given (X, Y) is particularly direct for scalar Q. It basically proceeds in three steps: first, average over multiple imputations given (X, Y, I, R), assuming the asymptotic validity of the sampling distribution of S_m; second, average over R given (X, Y, I), assuming the imputation method is proper under the posited response mechanism; and finally, average over I given (X, Y), assuming the complete-data inference is randomization valid under the specified unconfounded sampling mechanism.

From (4.5.1) and (4.5.2), the first step gives

$$(\overline{Q}_m | X, Y, I, R) \sim N(\overline{Q}_\infty, B_\infty/m), \qquad (4.5.3)$$

$$(\overline{U}_m | X, Y, I, R) \sim (\overline{U}_\infty, \ll B_\infty/m), \qquad (4.5.4)$$

and

$$((m-1)B_m B_\infty^{-1} | X, Y, I, R) \sim \chi^2_{m-1}, \qquad (4.5.5)$$

where these three random variables are mutually independent given (X, Y, I, R).

Now perform step 2. Expressions (4.2.5), (4.2.6) (4.5.3), and (4.5.5) imply that

$$(\overline{Q}_m | X, Y, I) \sim N(\hat{Q}, (1 + m^{-1})B) \qquad (4.5.6)$$

and

$$((m-1)B_m B^{-1} | X, Y, I) \sim \chi^2_{m-1}; \qquad (4.5.7)$$

expressions (4.2.8), (4.2.6), and (4.5.4) imply that

$$(\overline{U}_m | X, Y, I) \sim (U, \ll (1 + m^{-1})B). \qquad (4.5.8)$$

The three random variables in (4.5.6), (4.5.7), and (4.5.8) are mutually independent given (X, Y, I).

Finally, perform step 3. Expressions (4.2.3), (4.2.4), (4.2.9), and (4.5.6) imply that

$$(\overline{Q}_m | X, Y) \sim N(Q, U_0 + (1 + m^{-1})B_0); \quad (4.5.9)$$

expressions (4.5.9) and (4.5.7) imply that

$$((m - 1)B_m B_0^{-1} | X, Y) \sim \chi^2_{m-1}; \quad (4.5.10)$$

expressions (4.2.4), (4.2.9), and (4.5.8) imply that

$$(\overline{U}_m | X, Y) \sim (U_0, \ll (U_0 + (1 + m^{-1})B_0)); \quad (4.5.11)$$

the three random variables in (4.5.9), (4.5.10), and (4.5.11) are mutually independent given (X, Y).

Random-Response Randomization-Based Justification for the t Reference Distribution

Expressions (4.5.9), (4.5.10), and (4.5.11), which comprise the asymptotic sampling distribution of S_m given (X, Y) for proper imputation methods, imply that

$$E(T_m | X, Y) = V(\overline{Q}_m | X, Y) \quad (4.5.12)$$

and

$$V(T_m | X, Y) = 2(1 + m^{-1})^2 B_0^2 / (m - 1), \quad (4.5.13)$$

which together with (4.5.9) imply that

$$2E(T_m | X, Y)^2 / V(T_m | X, Y)$$
$$= (m - 1)\{1 + [(1 + m^{-1})B_0 / U_0]^{-1}\}^2. \quad (4.5.14)$$

Expressions (4.5.9), (4.5.12), and (4.5.14) provide a random-response randomization-based justification for using a t reference distribution for $(\overline{Q}_m - Q)$ with squared scale T_m and $\nu = (m - 1)(1 + r_m^{-1})^2$ degrees of freedom–note: $r_m = (1 + m^{-1})B_m / \overline{U}_m$, which estimates $(1 + m^{-1})B_0 / U_0$. The reason is that for any multiple of a χ^2 random variable, twice the squared expectation divided by the variance gives the degrees of freedom.

Extension of Results to Multicomponent Q

It is easy to see that the results already derived for scalar Q provide most of the results needed for multicomponent Q: (4.5.9) for $\bar{Q}_m$ and (4.5.11) for $\bar{U}_m$ hold for multicomponent Q, and (4.5.10) for B_m needs to be modified from the χ^2 to the k-dimensional Wishart distribution with $m-1$ degrees of freedom. Thus, in general,

$$(\bar{Q}_m | X, Y) \sim N(Q, U_0 + (1 + m^{-1})B_0); \quad (4.5.15)$$

$$((m-1)B_m | X, Y) \sim \text{Wishart}_k(m-1, B_0) \quad (4.5.16)$$

or equivalently,

$$(m-1)B_m = \sum_{i=1}^{m-1} Z_i Z_i' \quad \text{where} \quad Z_i \sim N(0, B_0);$$

and

$$(\bar{U}_m | X, Y) \sim (U_0, \ll (U_0 + (1 + m^{-1})B_0)); \quad (4.5.17)$$

where the three random variables $\bar{Q}_m$, B_m, and $\bar{U}_m$ are mutually independent.

Asymptotic Efficiency of $\bar{Q}_m$ Relative to $\bar{Q}_\infty$

From (4.5.15) the variance–covariance matrix of $\bar{Q}_m$ is $U_0 + (1 + m^{-1})B_0$, and so the variance–covariance matrix of $\bar{Q}_m$ relative to that of $\bar{Q}_\infty$ is

$$T_0^{-1/2}(T_0 + B_0/m)T_0^{-1/2}, \quad (4.5.18)$$

where

$$T_0 = U_0 + B_0. \quad (4.5.19)$$

Thus the efficiency of $\bar{Q}_m$ relative to $\bar{Q}_\infty$ in units of standard deviations is, from (4.5.18) and simple matrix algebra,

$$\text{RE} = (I + \gamma_0/m)^{-1/2}, \quad (4.5.20)$$

where

$$\gamma_0 = T_0^{-1/2} B_0 T_0^{-1/2} \quad (4.5.21)$$

is the diagonal matrix giving the population fractions of missing information—the eigenvalues of B_0 with respect to T_0, which from (4.5.19) lie between 0 and 1. Hence, these fractions of missing information determine the relative efficiency of estimation using a finite rather than infinite number of imputations. The largest fraction corresponds to the lowest relative efficiency for a linear combination of the components of Q, and the smallest fraction corresponds to the highest relative efficiency for such a linear combination. From (4.5.19) the eigenvalues of B_0 with respect to T_0 are one minus the eigenvalues of U_0 with respect to T_0. The smallest of these appears in the literature on missing data as the large-sample rate of convergence of the EM algorithm, an ubiquitous algorithm for maximum-likelihood estimation from incomplete data (Dempster, Laird, and Rubin, 1977).

4.6. EVALUATIONS OF FINITE-m INFERENCES WITH SCALAR ESTIMANDS

The preceding results can be easily applied to evaluate finite-m inferences with scalar Q. For instance, $\overline{Q}_m$ created using proper imputation methods has asymptotic relative efficiency (in units of standard deviations) given by (4.5.20) where $\gamma_0 = B_0/T_0$ is the population fraction of information about Q missing due to nonresponse. Values of RE appear in Table 4.1.

Small-Sample Efficiencies of Asymptotically Proper Imputation Methods from Examples 4.2–4.5

Consider the simple situation of Section 4.3 and the asymptotically proper imputation methods of Examples 4.2–4.5, but drop the restriction that the sample size is large. It is easy to see that for these methods

$$\overline{Q}_\infty = \bar{y}_1$$

and

$$V(\overline{Q}_m | X, Y, R, I) = B_\infty/m = (n_1^{-1} - n^{-1}) E(s_0^2 | X, Y, R, I)/m$$

where s_0^2 is the sample variance among the n_0 imputed values at an imputation. Thus, the random-response randomization-based variance of $\overline{Q}_\infty$ given unconfounded nonresponse and fixed n_1 is

$$V(\overline{Q}_\infty | X, Y, n_1) = V(\bar{y}_1 | X, Y, n_1) = S^2(n_1^{-1} - N^{-1}) \quad (4.6.1)$$

EVALUATION OF FINITE-m INFERENCES WITH SCALAR Q

and of $\overline{Q}_m$ is

$$
\begin{aligned}
V(\overline{Q}_m | X, Y, n_1) &= V\left[E(\overline{Q}_m | X, Y, R, I) | X, Y, n_1\right] \\
&\quad + E\left[V(\overline{Q}_m | X, Y, R, I) | X, Y, n_1\right] \\
&= V(\overline{Q}_\infty | X, Y, n_1) + E(B_\infty | X, Y, n_1)/m \\
&= S^2(n_1^{-1} - N^{-1}) + (n_1^{-1} - n^{-1}) E(s_0^2 | X, Y, n_1)/m.
\end{aligned}
$$
(4.6.2)

Hence, the small-sample efficiency of $\overline{Q}_m$ relative to $\overline{Q}_\infty$ with fixed n_1 is given by the square root of the ratio of (4.6.1) to (4.6.2), or after some algebraic manipulation,

$$\text{RE}' = (1 + c\gamma_0'/m)^{-1/2}, \tag{4.6.3}$$

where γ_0' is the fixed-n_1 fraction of missing information, $1 - V(\hat{Q} | X, Y, n_1)/V(\overline{Q}_\infty | X, Y, n_1)$, and $c = E(s_0^2/S^2 | X, Y, n_1)$. Straightforward algebra gives

$$\gamma_0' = \frac{n_1^{-1} - n^{-1}}{n_1^{-1} - N^{-1}} \doteq 1 - \frac{n_1}{n} \quad \text{for large } N,$$

and

$$
c = \begin{cases}
\dfrac{n_1 - 1}{n_1 + 1} & \text{for the Bayesian bootstrap (BB),} \\[1ex]
\dfrac{n_1 - 1}{n_1 - 3} & \text{for the fully normal and mean and variance adjusted hot-deck (FN and MV),} \\[1ex]
\left(\dfrac{n_1 - 1}{n_1}\right)^2 & \text{for the approximate Bayesian bootstrap (ABB).}
\end{cases}
$$

Thus, the small-sample relative efficiencies for these repeated imputation procedures are all very close to their asymptotic values given by the general result (4.5.20), as long as n_1 is not very small. Similar efficiency calculations appear in Rubin (1979a) and Herzog and Rubin (1983), and related working calculations are given in the appendix to Rubin and Schenker (1986) and in Schenker (1985).

Large-Sample Coverages of Interval Estimates Using a t Reference Distribution and Proper Imputation Methods

From (3.3.16), the $100(1 - \alpha)\%$ interval estimate of Q based on the t reference distribution is $C = \overline{Q}_m \pm t_\nu(\alpha/2) T_m^{1/2}$. The random-response randomization-based coverage of C for Q is thus

$$\text{Prob}\{Q \in C | X, Y\} = \text{Prob}\{|Q - \overline{Q}_m| < t_\nu(\alpha/2) T_m^{1/2} | X, Y\}.$$

The asymptotic sampling distribution of S_m given by (4.5.9)–(4.5.11) implies that

$$\overline{U}_m = U_0,$$

$$(m - 1) B_m / B_0 \equiv q \sim \chi^2_{m-1};$$

$$(\overline{Q}_m - Q)/[U_0 + (1 + m^{-1}) B_0]^{1/2} \equiv z \sim N(0,1),$$

with q and z independent. Thus

$$\text{Prob}\{Q \in C | X, Y\} = \text{Prob}\left\{ z^2 < t_\nu(\alpha/2)^2 \left(\frac{U_0 + (1 + m^{-1}) \dfrac{B_0 q}{m-1}}{U_0 + (1 + m^{-1}) B_0} \right) \right\}$$

with

$$\nu = (m - 1) \left[1 + (1 + m^{-1})^{-1} \frac{U_0(m-1)}{B_0 q} \right]^2.$$

Or letting $\rho_0 = B_0/U_0 = \gamma_0/(1 - \gamma_0)$,

$$\text{Prob}\{Q \in C | X, Y\} = \text{Prob}\left\{ z^2 < t_\nu(\alpha/2)^2 \left(\frac{1 + (1 + m^{-1}) \rho_0 \dfrac{q}{m-1}}{1 + (1 + m^{-1}) \rho_0} \right) \right\}$$

(4.6.4)

with

$$\nu = (m - 1) \left[1 + (1 + m^{-1})^{-1} \rho_0^{-1} \frac{m-1}{q} \right]^2.$$

EVALUATION OF FINITE-m INFERENCES WITH SCALAR Q

Consequently, without loss of generality when evaluating t-based inferences in large samples with asymptotically proper imputation methods, we can let $Q = 0$, $U_0 = 1$, and $B_0 = \rho_0 = \gamma_0/(1 - \gamma_0)$, and evaluate the coverage as a function of the number of imputations, m, and the fraction of missing information, γ_0. Thus, general advice on the coverage properties of multiple-imputation interval estimates is relatively straightforward to formulate.

Values of (4.6.4) can be found for various values of m and γ_0 by two-dimensional numerical integration or Monte Carlo methods. In Rubin and Schenker (1986), numerical integration was used to evaluate (4.6.4) as well as the corresponding probabilities for interval estimates based on the normal reference distribution [i.e., (4.6.4) with ν fixed at ∞]. Some results for the t reference distribution are given in Table 4.2; results for the normal reference distribution are uniformly inferior and so are not tabulated here except for the case $m = 1$ when the t reference distribution is not defined.

To obtain approximate coverages for values of $m > 2$ not given in Table 4.2, work reported in Rubin and Schenker (1986) suggests that linear interpolation in $(m - 1)^{-1}$ gives reasonably accurate answers (remember that when $m = \infty$, the coverages equal the nominal levels because proper imputation methods are randomization-valid).

Small-Sample Monte Carlo Coverages of Asymptotically Proper Imputation Methods from Examples 4.2–4.5

We now summarize a Monte Carlo study designed to investigate the asymptotic propriety of the asymptotic results given by (4.6.4) and displayed in Table 4.2. Rubin and Schenker (1986) applied the asymptotically proper repeated-imputation procedures described in Examples 4.2–4.5 to data generated from the normal, Laplace, and lognormal distributions (all with $N = \infty$). They then calculated the actual frequency coverages of the resulting t-based intervals for $\bar{Y}$ using the standard complete-data statistics $\bar{y}$ and s^2/n with fixed response rate $n_1/n = 1 - \gamma_0'$. Table 4.3 summarizes the results for $m = 2$. Clearly the asymptotic results are approached most rapidly with normal data and least rapidly with lognormal data; the discrepancies from nominal levels are due in large part to the propriety of the standard complete-data inference with small samples when the observations are not normal.

Evaluation of Significance Levels

Accuracy of significance levels using the multiple-imputation procedures described in this section can be found using Tables 4.2 and 4.3 for confidence coverages because of the exact correspondence between con-

TABLE 4.3. Simulated coverages (in %) of asymptotically proper multiple ($m = 2$) imputation procedures with nominal levels 90% and 95%, using t-based inferences, response rates n_1/n, and normal and nonnormal data (Laplace, lognormal = $\exp N(0, 1)$); maximum standard error < 1%.[a]

| | | Normal | | | | | | | | Laplace | | | | | | | | Lognormal (exp. of unit normal) | | | | | | | |
| | | 90% | | | | 95% | | | | 90% | | | | 95% | | | | 90% | | | | 95% | | | |
n	n_1/n	BB	ABB	FN	MV	BB	ABB	FN	MV	BB	ABB	FN	MV	BB	ABB	FN	MV	BB	ABB	FN	MV	BB	ABB	FN	MV
20	.4	81	82	86	86	87	87	90	91	81	81	85	85	87	87	90	90	73	73	78	78	79	79	83	83
	.6	85	85	87	86	90	90	92	91	85	85	88	87	90	91	93	92	77	77	79	79	82	82	85	84
	.9	88	88	89	89	93	93	94	94	89	88	89	89	94	94	94	95	80	80	80	80	84	84	85	84
50	.4	84	84	86	85	89	89	91	90	86	83	84	85	90	89	88	90	79	77	81	80	84	83	86	85
	.6	86	87	88	87	92	92	93	92	86	87	86	86	91	92	91	92	82	81	82	83	87	87	87	88
	.9	89	89	90	89	94	94	95	94	88	88	88	88	94	94	94	94	85	85	85	85	90	90	90	90
100	.4	85	85	85	86	90	90	90	90	84	83	86	85	89	89	90	90	82	81	81	82	87	87	87	87
	.6	87	87	88	88	92	92	93	93	87	87	88	87	92	92	92	92	84	84	85	84	89	89	89	89
	.9	90	90	90	90	95	95	95	95	89	89	89	89	94	95	94	94	86	86	87	86	92	91	92	91
1000	.4	90	90	90	90	95	95	95	95	86	85	84	86	91	90	89	90	85	83	86	85	90	88	90	89
	.6	90	90	90	90	95	95	95	95	89	87	87	88	94	93	92	92	87	87	87	88	91	92	92	93
	.9	90	90	90	90	95	95	95	95	91	90	90	90	95	95	95	95	90	90	90	90	95	95	95	95

BB: Bayesian Bootstrap–Example 4.3
ABB: Approximate Bayesian Bootstrap–Example 4.4
FN: Fully Normal method–Examples 3.2 and 4.2
MV: Mean and Variance adjusted hot-deck–Example 4.5
[a]Source: Rubin and Schenker (1986).

fidence intervals and tests using the same statistics: $100(1 - \alpha)\%$ confidence intervals correspond to $100\alpha\%$ level tests. Thus, for example, from Table 4.2, when $\gamma_0 = .4$, $m = 2$ and $\alpha = 5\%$, the actual level is one minus the displayed value, or 7%. For a small-sample example, with the same values of $\gamma'_0 = 1 - n_1/n$, m, and α, and using the Bayesian bootstrap with 100 observations from the Laplace distribution, from Table 4.3 the actual level for a 5% test is $100 - 92 = 8\%$.

When the imputation method is proper and $m = \infty$, the actual levels of tests will equal their nominal levels since the resulting inferences are randomization-valid. Of course, in practice, the important role of significance levels is to summarize evidence about multicomponent rather than scalar estimands.

4.7. EVALUATION OF SIGNIFICANCE LEVELS FROM THE MOMENT-BASED STATISTICS D_m AND $\tilde{D}_m$ WITH MULTICOMPONENT ESTIMANDS

Several procedures have been suggested in Chapter 3 for finding the significance level associated with the null value Q_0 when Q has more than one component. Both D_m and $\tilde{D}_m$ are simple functions of $(\bar{Q}_m, \bar{U}_m, B_m)$:

$$D_m = (Q_0 - \bar{Q}_m)\left[\bar{U}_m + (1 + m^{-1})B_m\right]^{-1}(Q_0 - \bar{Q}_m)^t \quad (4.7.1)$$

$$\tilde{D}_m = (1 + r_m)^{-1}(Q_0 - \bar{Q}_m)\bar{U}_m^{-1}(Q_0 - \bar{Q}_m)^t, \quad (4.7.2)$$

where

$$r_m = (1 + m^{-1})\operatorname{Tr}(B_m \bar{U}_m^{-1})/k. \quad (4.7.3)$$

The suggested reference distributions for finding p-values are both F distributions

$$p\text{-value}(Q_0|D_m) = \operatorname{Prob}\{F_{k,\nu} > D_m\} \quad (4.7.4)$$

$$p\text{-value}(Q_0|\tilde{D}_m) = \operatorname{Prob}\{F_{k,(k+1)\nu/2} > \tilde{D}_m\}, \quad (4.7.5)$$

where

$$\nu = (m - 1)(1 + r_m^{-1})^2. \quad (4.7.6)$$

The Level of a Significance Testing Procedure

From the random-response randomization-based perspective, under the null hypothesis that $Q = Q_0$, p-values should have a uniform distribution given (X, Y), averaging over (R, I) and the set of multiple imputations. Specifically, consider the statistic D (e.g., D_m or $\tilde{D}_m$), the null value $Q = Q_0$, and the 100α percentage point of the reference distribution, $F_{k, k'\nu}(\alpha)$, where k' is 1 for D_m and $(k+1)/2$ for $\tilde{D}_m$. Then the procedure is said to have the correct level α if

$$\text{Prob}\{D > F_{k, k'\nu}(\alpha) | X, Y; Q = Q_0\} \quad (4.7.7)$$

equals α. Nominal values of α that tend to be of particular common interest include 10%, 5% and 1%. Considerations of power, such as the loss of power due to finite m, are closely related to issues of loss of efficiency due to finite m, and are also relevant, but are secondary to the primary issue of having procedures with approximately valid nominal levels. Work on the evaluation of p-values is at an early stage of development so that the results presented here are limited to finding the levels of proper imputation procedures in large samples; thus, the joint distribution of $(\bar{Q}_m, \bar{U}_m, B_m)$ given in Section 4.5 by (4.5.15)–(4.5.17) is accepted.

The Level of D_m—Analysis for Proper Imputation Methods and Large Samples

The actual level of the nominal α level procedure that uses the statistic D_m and the reference distribution $F_{k,\nu}$ is given by (4.7.7) with $D = D_m$ and $k' = 1$. For purposes of general evaluations, we assume proper imputation methods [(4.2.5)–(4.2.10)], valid complete-data inferences [(4.2.3) and (4.2.4)], and the validity of the asymptotic sampling distribution of the (Q_{*l}, U_{*l}) given by (4.5.1) and (4.5.2). Straightforward manipulations then show that this probability is identical to

$$\text{Prob}\{Z_0[I + (1 + m^{-1})W]^{-1}Z_0' > kF_{k,\nu}(\alpha)\}, \quad (4.7.8)$$

where

$$\nu = (m-1)\left[1 + (1 + m^{-1})^{-1}k/\text{Tr}(W)\right]^2,$$

$$Z_0 \sim N(0, I + (1 + m^{-1})\rho_0),$$

$$W = \sum_{i=1}^{m-1} Z_i' Z_i / (m-1),$$

$$Z_i \stackrel{\text{i.i.d.}}{\sim} N(0, \rho_0) \text{ independent of } Z_0,$$

SIGNIFICANCE LEVELS FROM MOMENT-BASED STATISTICS

and

ρ_0 is the $k \times k$ diagonal matrix of the eigenvalues of B_0 with respect to U_0, $\rho_0 = \text{diag}(\rho_{01},\ldots,\rho_{0k})$.

Expression (4.7.8) is easily evaluated by Monte Carlo methods as a function of α, m, k, and ρ_0 by drawing m k-variate normal random variables at each replication. More efficient methods for evaluating the level of D_m exist, especially when m is large; Li (1985) provides some details. Incidentally, power calculations can be performed by having Z_0 centered at the alternative δ rather than 0.

The Level of D_m—Numerical Results

Inspection of (4.7.8) suggests that the level of D_m may be relatively insensitive to variability in the eigenvalues $\rho_{01},\ldots,\rho_{0k}$, which are equivalent to the population fractions of information missing due to nonresponse, $\gamma_{01},\ldots,\gamma_{0k}$, $\gamma_0 = \rho_0(I + \rho_0)^{-1}$. This conclusion is supported by the numerical results, and so only results with γ_0 proportional to the identity matrix will be presented. Also if m is large enough, it is clear from (4.7.8) that the level of D_m will be accurate, so the essential question with D_m is what value of m constitutes "large enough." Table 4.4, based on the work of Li (1985), suggests that as a rough rule, the level of D_m is accurate when $m > 10\gamma_0 k$. Since 50% missing information is an extreme situation, this rule suggests that D_m will be accurate if $m \geq 5k$, as implied in Section 3.1.

The Level of $\tilde{D}_m$—Analysis

The actual level of the nominal α-level procedure that uses the statistic $\tilde{D}_m$ and the reference distribution $F_{k,k'\nu}$ is given by (4.7.7) with $D = \tilde{D}_m$. Straightforward manipulations show that the probability can be written in the notation of (4.7.8) as

$$\text{Prob}\{[k + \text{Tr}(W)]^{-1}Z_0 Z_0^t > F_{k,k'\nu}(\alpha)\}, \qquad (4.7.9)$$

where

$$Z_0 Z_0^t = \sum_{i=1}^{k}\left[1 + (1 + m^{-1})\rho_{0i}\right]\chi_{i,1}^2, \qquad \chi_{1,i}^2 \overset{\text{i.i.d.}}{\sim} \chi_1^2 \qquad (4.7.10)$$

$$\text{Tr}(W) = \sum_{i=1}^{k}\rho_{0i}\chi_{m-1,i}^2/(m-1), \qquad \chi_{m-1,i}^2 \overset{\text{i.i.d.}}{\sim} \chi_{m-1}^2, \qquad (4.7.11)$$

with the $2k$ χ^2 random variables mutually independent.

TABLE 4.4. Large-sample level (in %) of D_m with $F_{k,\nu}$ reference distribution as a function of nominal level, α; number of components being tested, k; number of proper imputations, m; and fraction of missing information, γ_0. Accuracy of results = 5000 simulations of (4.7.8) with ρ_0 set to I.

			$\alpha = 1\%$				$\alpha = 5\%$				$\alpha = 10\%$			
k	m	$\gamma_0 \doteq$	.1	.2	.3	.5	.1	.2	.3	.5	.1	.2	.3	.5
2	2		1.0	1.2	1.6	2.5	4.9	5.3	5.9	7.5	9.9	10.3	11.0	12.9
	3		1.0	1.0	1.0	1.3	4.9	4.9	5.0	5.5	9.9	9.8	10.0	10.9
	5		1.0	1.0	1.1	1.2	5.0	5.0	5.1	5.6	10.0	10.0	10.2	10.9
	10		1.0	1.0	1.1	1.2	5.0	5.1	5.3	5.7	10.1	10.2	10.4	11.0
	25		1.0	1.0	1.0	1.0	5.0	5.0	5.0	5.0	10.0	9.9	9.9	10.0
	50		1.0	1.0	1.0	1.0	5.0	5.0	5.0	5.0	10.0	9.9	9.9	10.0
	100		1.0	1.0	1.0	1.0	5.0	5.0	5.0	5.0	10.0	10.0	10.0	10.1
3	2		1.0	1.1	1.3	1.7	5.1	5.3	5.6	6.3	10.3	10.6	11.1	12.0
	3		1.0	1.0	1.0	1.0	5.1	5.2	5.3	5.7	10.2	10.5	10.9	12.3
	5		1.0	1.0	1.1	1.3	5.0	5.2	5.4	6.2	10.1	10.3	10.8	12.2
	10		1.0	1.0	1.1	1.2	5.0	5.2	5.3	5.9	10.1	10.3	10.6	11.6
	25		1.0	1.0	1.1	1.2	5.0	5.1	5.2	5.6	10.1	10.2	10.4	10.9
	50		1.0	1.0	1.0	1.0	5.0	5.0	5.0	5.1	10.0	10.0	10.0	10.2
	100		1.0	1.0	1.0	1.0	5.0	5.0	5.1	5.1	10.0	10.0	10.1	10.2
5	2		0.9	0.8	0.8	0.9	5.1	4.8	4.5	4.0	10.5	10.4	10.1	9.2
	3		1.0	1.0	1.0	0.9	5.2	5.5	5.7	6.1	10.5	11.3	12.1	14.4
	5		1.1	1.1	1.2	1.4	5.2	5.6	6.1	7.7	10.4	11.1	12.2	15.4
	10		1.0	1.1	1.2	1.5	5.1	5.3	5.6	6.9	10.1	10.4	11.1	13.1
	25		1.0	1.0	1.1	1.3	5.0	5.2	5.3	6.0	10.1	10.3	10.6	11.5
	50		1.0	1.0	1.0	1.1	5.0	5.1	5.1	5.4	10.0	10.1	10.2	10.7
	100		1.0	1.0	1.0	1.1	5.0	5.0	5.1	5.2	10.0	10.1	10.1	10.4
10	2		0.8	0.5	0.3	0.1	5.1	4.0	2.9	1.5	10.8	10.1	8.5	5.4
	3		1.1	0.9	0.6	0.3	5.6	5.9	5.7	4.9	11.3	12.7	13.8	16.2
	5		1.1	1.2	1.3	1.4	5.4	6.3	7.4	11.0	10.7	12.4	14.8	22.7
	10		1.1	1.2	1.4	2.2	5.2	5.8	6.8	10.3	10.4	11.4	13.1	19.0
	25		1.0	1.1	1.2	1.6	5.0	5.2	5.6	7.1	10.0	10.4	11.0	13.4
	50		1.0	1.0	1.1	1.3	5.0	5.1	5.4	6.1	10.0	10.2	10.6	11.8
	100		1.0	1.0	1.1	1.2	5.0	5.2	5.3	5.8	10.1	10.2	10.5	11.3

Expression (4.7.9) can be easily evaluated by Monte Carlo methods using k independent χ_1^2 and k independent χ_{m-1}^2 random variables at each repetition; these can be calculated from the same normal deviates used to generate the $Z_0, \ldots, Z_{m-1}$ in (4.7.8) when both D_m and $\tilde{D}_m$ are being evaluated in one simulation. If some of the eigenvalues ρ_{0i} are equal, the number of χ^2 random variables can be reduced. For example, in the

SIGNIFICANCE LEVELS FROM MOMENT-BASED STATISTICS

extreme case when all $\rho_{0i} = \rho_0.$,

$$Z_0 Z_0^t = [1 + (1 + m^{-1})\rho_0.]\chi_k^2$$

and

$$\text{Tr}(W) = \frac{\rho_0.}{m-1}\chi_{k(m-1)}^2,$$

and then (4.7.9) can be easily evaluated by two-dimensional numerical integration since it equals

$$\text{Prob}\left\{ \frac{[1 + (1 + m^{-1})\rho_0.]\chi_k^2/k}{1 + \frac{\rho_0.}{k(m-1)}\chi_{k(m-1)}^2} > F_{k,k'\nu}(\alpha) \right\}. \quad (4.7.12)$$

Incidentally, power calculations can be performed by having $\chi_{1,i}^2$ in (4.7.10) and χ_k^2 in (4.7.12) replaced by noncentral χ^2 random variables with noncentrality parameters δ_i^2 and $\Sigma\delta_i^2$, respectively, where $\delta = (\delta_1, \ldots, \delta_k)$.

The Effect of Unequal Fractions of Missing Information on $\tilde{D}_m$

An important question concerns how (4.7.9) compares with (4.7.12); that is, what is the effect on $\tilde{D}_m$ of unequal eigenvalues? Recall that 50% missing information implies $\rho_{0i} = 1$ and no missing information implies that $\rho_{0i} = 0.0$.

Notice from (4.7.10) that unless some of the ρ_{0i} are very large, the coefficients of $\chi_{1,i}^2$ will be nearly equal, suggesting that even if a Satterthwaite (1946) type approximation were used for $Z_0 Z_0^t$, the degrees of freedom for $Z_0 Z_0^t$ would usually still be close to k, the value when all ρ_{0i} are equal. In contrast, notice from (4.7.11) that if all but one of the ρ_{0i} are very small, the degrees of freedom for $\text{Tr}(W)$ will be closer to $(m-1)$ than $k(m-1)$. Hence, whereas in most cases the numerator degrees of freedom of $\tilde{D}_m$ is reasonably close to k, an appropriate denominator degrees of freedom can vary between ν and $k\nu$ depending on the fractions of missing information; $(k+1)\nu/2$ is midway between these extremes.

Some Numerical Results for $\tilde{D}_m$ with $k' = (k+1)\nu/2$

Table 4.5 gives the actual level of $\tilde{D}_m$ using the $F_{k,(k+1)\nu/2}$ reference distribution for nominal levels $\alpha = 1, 5,$ and 10%, respectively in parts a, b, c.

TABLE 4.5. Large-sample level (in %) of $\tilde{D}_m$ with $F_{k,(k+1)\nu/2}$ reference distribution as a function of number of components being tested, k; number of proper imputations, m; fraction of missing information, γ_0; and variance of fractions of missing information, 0 (zero), S (small), L (large). Accuracy of results = 5000 simulations of (4.7.9).

k	m	$\gamma_0 = .1$			$\gamma_0 = .2$			$\gamma_0 = .3$			$\gamma_0 = .5$		
		0	S	L	0	S	L	0	S	L	0	S	L
(a) Nominal level $\alpha = 1\%$													
2	2	0.7	0.9	0.7	0.9	1.2	0.9	1.1	1.5	1.2	2.0	2.4	2.2
	3	0.7	1.0	0.7	0.7	1.1	0.8	0.8	1.2	0.9	1.1	1.9	1.3
	5	0.7	1.1	0.7	0.7	1.0	0.8	0.7	1.0	0.8	0.8	1.5	0.9
	10	0.7	1.1	0.7	0.7	1.0	0.7	0.7	1.0	0.7	0.7	1.1	0.8
3	2	0.9	0.9	0.9	1.0	1.0	1.1	1.2	1.3	1.3	1.9	1.5	2.1
	3	0.9	0.9	1.0	0.9	0.9	1.0	0.9	1.0	1.0	1.1	1.1	1.2
	5	0.9	0.9	1.0	0.9	0.9	0.9	0.9	0.9	0.9	0.9	1.0	1.0
	10	0.9	1.0	1.0	0.9	0.9	0.9	0.9	0.9	0.9	0.8	1.0	0.9
5	2	0.9	1.0	1.1	1.0	1.0	1.2	1.1	1.2	1.3	1.5	1.1	1.7
	3	1.0	1.0	1.1	0.9	1.0	1.0	0.9	1.0	1.0	0.9	0.9	1.1
	5	1.0	1.0	1.1	0.9	1.0	1.0	0.9	0.9	1.0	0.8	0.9	1.0
	10	1.0	1.1	1.1	1.0	1.0	1.0	0.9	1.0	1.0	0.9	0.9	1.0
10	2	0.9	1.0	1.2	0.9	1.0	1.1	0.9	1.0	1.2	1.0	1.0	1.2
	3	1.0	1.0	1.1	0.9	1.0	1.1	0.9	0.9	1.0	0.8	0.9	1.0
	5	1.0	1.1	1.1	1.0	1.0	1.1	0.9	0.9	1.0	0.8	0.9	1.0
	10	1.0	1.2	1.2	1.0	1.0	1.1	1.0	1.0	1.1	0.9	0.9	1.1
(b) Nominal level $\alpha = 5\%$													
2	2	4.7	4.3	4.3	4.9	4.7	5.0	5.4	5.3	5.5	6.8	6.7	7.0
	3	4.7	4.7	4.6	4.7	4.7	4.9	4.8	4.9	5.0	5.2	5.3	5.5
	5	4.7	4.8	4.8	4.7	4.7	4.8	4.7	4.7	4.8	4.7	4.8	4.9
	10	4.7	4.9	4.9	4.7	4.7	4.8	4.7	4.7	4.8	4.6	4.7	4.8
3	2	4.8	4.7	4.7	4.9	4.9	5.1	5.2	5.3	5.4	6.1	6.2	6.4
	3	4.9	4.9	5.0	4.8	4.9	5.0	4.8	4.8	5.0	4.9	4.9	5.1
	5	4.9	5.0	5.1	4.8	4.9	4.9	4.8	4.8	4.9	4.6	4.7	4.8
	10	4.9	5.2	5.2	4.9	4.9	5.0	4.9	4.9	4.9	4.7	4.8	4.9
5	2	4.8	4.9	5.0	4.8	4.9	5.1	4.9	5.0	5.2	5.2	5.3	5.6
	3	4.9	5.1	5.2	4.9	4.8	5.0	4.7	4.7	4.9	4.5	4.6	4.8
	5	5.0	5.2	5.2	4.9	4.9	5.0	4.8	4.8	4.9	4.5	4.6	4.7
	10	5.0	5.5	5.3	5.0	5.0	5.1	4.9	4.9	5.0	4.7	4.8	4.9
10	2	4.8	5.1	5.4	4.6	4.8	5.2	4.5	4.6	5.0	4.3	5.2	4.8
	3	4.9	5.2	5.3	4.9	4.8	5.1	4.5	4.6	4.9	4.1	4.2	4.7
	5	5.0	5.4	5.3	4.9	5.0	5.1	4.7	4.8	5.0	4.4	4.3	4.9
	10	5.1	5.9	5.5	5.0	5.1	5.2	4.9	4.9	5.1	4.7	4.8	5.2

TABLE 4.5. (*Continued*)

k	m	$\gamma_0 = .1$			$\gamma_0 = .2$			$\gamma_0 = .3$			$\gamma_0 = .5$		
		0	S	L	0	S	L	0	S	L	0	S	L
					(c) Nominal level $\alpha = 10\%$								
2	2	9.6	8.9	8.6	9.8	9.4	9.7	10.2	10.0	10.3	11.5	11.4	11.7
	3	9.7	9.6	9.4	9.7	9.7	9.9	9.7	9.8	10.0	9.9	10.1	10.3
	5	9.8	10.0	9.9	9.7	9.8	9.9	9.7	9.7	9.8	9.6	9.7	9.9
	10	9.8	10.2	10.1	9.8	9.8	9.9	9.8	9.8	9.8	9.7	9.7	9.8
3	2	9.7	9.4	9.3	9.7	9.6	9.9	9.9	9.9	10.2	10.6	10.7	11.0
	3	9.9	10.0	9.9	9.7	9.8	9.9	9.6	9.7	9.8	9.5	9.6	9.8
	5	9.9	10.2	10.2	9.8	9.9	9.9	9.7	9.8	9.8	9.5	9.5	9.7
	10	10.0	10.6	10.4	9.9	9.9	10.0	9.9	9.9	9.9	9.7	9.7	9.8
5	2	9.7	9.8	9.9	9.5	9.7	10.0	9.5	9.7	10.0	9.6	9.7	10.1
	3	9.9	10.1	10.2	9.7	9.8	9.9	9.5	9.6	9.7	9.0	9.2	9.4
	5	10.0	10.3	10.3	9.8	9.9	10.0	9.7	9.7	9.8	9.3	9.4	9.6
	10	10.0	11.1	10.6	9.9	10.0	10.1	9.9	9.9	10.0	9.6	9.7	9.8
10	2	9.7	10.1	10.5	9.3	9.6	10.0	9.1	9.3	9.8	8.5	9.0	9.4
	3	9.9	10.3	10.4	9.6	9.7	10.0	9.4	9.5	9.8	8.7	9.2	9.4
	5	10.0	10.8	10.5	9.8	9.9	10.1	9.6	9.7	10.0	9.2	9.5	9.8
	10	10.1	11.8	11.0	9.9	10.3	10.2	9.9	9.9	10.1	9.6	9.8	10.1

The factors in these tables that are known to the data analyst are m = the number of repeated imputations and k = the number of components being tested, and the factors that are unknown (although to some extent estimable from the data) are γ_0 = the average fraction of missing information and the variance of the fractions of missing information, where 0 = zero = no variance, S = small variance = 0.05^2, and L = large variance = 0.09^2 (the values are symmetrically distributed about the mean, γ_0; see Raghunathan, 1987 for the specific values chosen and for details of the numerical methods for computing values in the tables).

Results in these tables are extremely encouraging. If $\gamma_0 \leq 0.2$, even two repeated imputations appear to result in accurate levels, and three repeated imputations result in accurate levels even when $\gamma_0 = 0.5$. As k gets larger, the tests become more conservative when there is no variation in the fractions of missing information, which is expected since $k' = k$ rather than $k' = (k + 1)/2$ gives the proper reference distribution for large m. But for the more realistic cases with variation in the fractions of missing information (i.e., levels S and L) there is no obvious trend for the tests to become less accurate as k increases.

Nevertheless, improvements are possible because, if $m \geq 3$, there exists information in S_m about the variance of the fractions of missing information. Work in progress by Raghunathan suggests an adjustment based on the trace of $(B_m \bar{U}_m^{-1})^2$. Such an adjustment should also improve the power of the tests, as studied by Raghunathan (1987).

4.8. EVALUATION OF SIGNIFICANCE LEVELS BASED ON REPEATED SIGNIFICANCE LEVELS

Even though Table 4.5 suggests that $\tilde{D}_m$ using the $F_{k,(k+1)\nu/2}$ reference distribution can work well in practice, an important limitation remains: $\tilde{D}_m$ requires S_m, the set of complete-data moments. The asymptotically equivalent $\hat{D}_m$ requires first, $\{d_{*l}\}$, the set of $m\chi^2$ statistics, which are asymptotically equal to $\hat{Q}_{*l} \bar{U}_\infty^{-1} \hat{Q}'_{*l}$, and second, $r_m = \text{Tr}(B_m \bar{U}_m^{-1})$. If k is large, however, complete-data analyses will commonly summarize evidence using only the d_{*l} or the associated p-values, since the complete-data moments can require substantial allocation of storage and computational resources, and thus commonly with large k, r_m will not be available.

The Statistic $\hat{\hat{D}}_m$

Currently the only statistic that is a function solely of $\{d_{*l}\}$ and has been carefully studied is $\hat{\hat{D}}_m$ derived in Section 3.5 and given by

$$\hat{\hat{D}}_m = \frac{\dfrac{\bar{d}_m}{k} - \dfrac{m-1}{m+1}\hat{r}_m}{1 + \hat{r}_m},$$

where $\hat{r}_m$ is the method of moments estimator given by (3.5.8); the associated reference distribution is F on k and $(1 + k^{-1})\hat{\nu}/2$ degrees of freedom where $\hat{\nu}$ is given by (3.5.10). The two functions of the d_{*l} needed to calculate $\hat{\hat{D}}_m$ and its reference distribution are their mean, $\bar{d}_m$ and variance s_d^2.

The Asymptotic Sampling Distribution of $\bar{d}_m$ and s_d^2

As done previously, for general evaluation purposes we assume proper imputation methods [(4.2.5)–(4.2.10)], valid complete-data inferences [(4.2.3) and (4.2.4)], and the validity of the asymptotic sampling distribution of the (Q_{*l}, U_{*l}) given by (4.5.1) and (4.5.2). The resulting distribution of $(\bar{d}_m, s_d^2)$

SIGNIFICANCE LEVELS BASED ON REPEATED SIGNIFICANCE LEVELS

given (X, Y) is not easy to handle analytically, and so the evaluations of $\hat{D}_m$ that are presented are obtained via simulations. The quantities Q, U_0, B_0 are fixed, without loss of generality, at 0, the identity and $\rho_0 = \mathrm{diag}(\rho_{01},\ldots,\rho_{0k})$, respectively. A value of $\bar{Q}_\infty$ is drawn from $N(0, I + \rho_0)$; then m values of $\hat{Q}_{*l}$ are drawn from $N(\bar{Q}_\infty, \rho_0)$ to find $\bar{d}_m = \Sigma \hat{Q}_{*l} \hat{Q}^t_{*l}/m$ and $s_d^2 = \Sigma(\hat{Q}_{*l}\hat{Q}^t_{*l} - \bar{d}_m)^2/(m-1)$, and thereby $\hat{D}_m$ and its reference distribution. A new value of $\bar{Q}_\infty$ creates a new value of $\hat{D}_m$, and so on. Levels of $\hat{D}_m$ are estimated by the fraction of drawn values of $\hat{D}_m$ larger than $F_{k,(1+k^{-1})\hat{\nu}/2}(\alpha)$, that is, by the simulation analog of (4.7.7) with $D = \hat{D}_m$, $k' = (1 + k^{-1})/2$, and $\nu = \hat{\nu}$.

Some Numerical Results for $\hat{D}_m$

Table 4.6 gives the actual level of $\hat{D}_m$ using the $F_{k,(1+k^{-1})\hat{\nu}/2}$ reference distribution for nominal levels = 1, 5, and 10%, respectively in parts a, b, c, for the same factors as in Table 4.5. The results for $\hat{D}_m$ suggest that much work remains to be done except when it is known that γ_0 is small (e.g., $\leq 10\%$), or γ_0 is modest (e.g., $\leq 30\%$) and $m \geq k$. Fortunately, work in progress by Raghunathan (1987) promises improved procedures for use with $m \geq 3$.

The Superiority of Multiple-imputation Significance Levels

Even though the results in Table 4.6 for $\hat{D}_m$ leave room for major improvements in some situations, they are substantially better than the corresponding results with single imputation. In particular, with single proper imputation, the test statistic is d_{*1} with a χ^2_k reference distribution. It is easy to see that the level of this procedure, that is, expression (4.4.7) with $D = d_{*1}$ and $k' = \infty$, equals

$$\mathrm{Prob}\{Z_0 Z_0^t > \chi^2_k(\alpha)\}, \qquad (4.8.1)$$

where Z_0 is defined following both (4.7.8) and (4.7.9). Since results are relatively insensitive to variation in the fractions of missing information, (4.8.1) essentially equals $\mathrm{Prob}\{(1 + 2\rho_0.)\chi^2_k > \chi^2_k(\alpha)\}$, or equivalently

$$\mathrm{Prob}\left\{\left[1 + 2\gamma_0(1-\gamma_0)^{-1}\right]\chi^2_k > \chi^2_k(\alpha)\right\}. \qquad (4.8.2)$$

Even for modest γ_0 and modest k, (4.8.2) is substantially larger than the nominal level α; Table 4.7 displays results for the values of α, k, and γ_0

TABLE 4.6. Large-sample level (in %) of $\hat{D}_m$ with $F_{k,(1+k^{-1})\hat{\nu}/2}$ reference distribution as a function of number of components being tested, k; number of proper imputations, m; fraction of missing information, γ_0; and variance of fractions of missing information, 0 (zero), S (small), L (large). Accuracy of results = 5000 simulations of (4.7.7).

		$\gamma_0 = .1$			$\gamma_0 = .2$			$\gamma_0 = .3$			$\gamma_0 = .5$		
k	m	0	S	L	0	S	L	0	S	L	0	S	L
				(a) Nominal level $\alpha = 1\%$									
2	2	0.9	1.2	1.1	1.5	1.6	1.6	2.1	2.2	2.3	4.7	5.3	4.7
	3	1.0	1.1	1.1	1.2	1.2	1.1	1.5	1.4	1.5	2.9	3.1	2.6
	5	1.0	0.9	1.1	0.9	1.0	0.9	1.3	1.0	1.0	1.7	1.9	1.7
	10	0.9	1.0	1.1	1.0	1.0	0.9	1.0	0.9	0.9	1.4	1.2	0.9
3	2	1.2	1.3	1.2	1.7	1.8	1.9	2.6	2.8	2.5	6.4	7.0	6.9
	3	0.9	1.2	1.1	1.2	1.3	1.1	1.7	1.7	1.6	3.4	4.1	3.8
	5	0.9	1.0	1.1	1.0	1.0	1.0	1.3	1.3	1.1	2.2	2.3	2.2
	10	0.9	1.0	1.1	1.0	1.0	0.9	1.0	0.9	0.9	1.6	1.4	1.2
5	2	1.3	1.4	1.2	2.2	2.3	2.4	3.6	3.7	3.4	9.8	10.5	10.3
	3	1.0	1.2	1.1	1.7	1.6	1.6	2.2	2.4	2.2	5.4	5.6	5.6
	5	1.1	1.0	1.0	1.3	1.2	1.2	1.4	1.5	1.3	2.7	3.1	2.9
	10	0.9	1.0	1.0	1.2	1.1	1.0	1.1	1.1	1.1	1.7	1.7	1.7
10	2	1.5	1.6	1.5	3.1	3.4	3.4	6.1	6.5	6.4	16.7	17.3	16.5
	3	1.3	1.3	1.1	2.0	2.1	2.1	3.7	3.6	3.5	10.0	9.7	8.8
	5	1.2	1.1	0.9	1.5	1.4	1.4	2.3	1.9	2.0	5.7	5.1	4.7
	10	1.0	1.0	0.8	1.2	1.2	1.0	1.5	1.1	1.5	2.6	2.6	1.9
				(b) Nominal level $\alpha = 5\%$									
2	2	4.7	4.8	4.7	5.4	6.0	5.5	7.3	7.4	7.1	11.3	11.9	11.0
	3	4.6	4.9	4.6	4.8	5.4	4.9	5.9	5.8	5.4	8.1	8.4	7.8
	5	4.8	4.9	4.6	4.7	5.1	4.6	5.2	5.0	4.5	6.5	6.6	6.1
	10	4.5	4.9	4.7	4.8	5.1	4.5	5.0	5.0	4.7	6.0	5.8	5.6
3	2	5.0	5.2	4.9	6.2	6.7	6.7	8.1	8.8	8.2	13.9	14.7	14.0
	3	4.8	5.0	4.9	5.5	5.7	5.3	6.4	6.6	6.4	10.2	10.0	9.5
	5	4.8	5.2	4.8	5.1	5.5	4.9	5.2	5.6	5.0	7.9	7.5	7.1
	10	4.9	5.1	4.7	5.1	5.3	4.6	4.8	5.1	4.6	6.2	5.8	5.9
5	2	5.5	5.7	5.4	7.2	7.7	8.1	10.5	10.9	10.1	19.0	19.4	18.7
	3	5.1	5.3	5.1	6.2	6.4	6.0	7.5	7.7	7.7	12.6	12.6	12.4
	5	4.9	5.3	4.8	5.4	5.5	5.1	6.3	6.2	6.0	8.8	9.7	8.7
	10	4.9	5.1	4.8	5.2	5.2	5.0	5.0	5.3	5.0	7.1	7.1	6.3
10	2	6.2	6.3	5.9	9.8	9.9	9.7	14.8	15.6	14.7	25.5	26.1	24.9
	3	5.5	5.4	5.4	7.8	7.9	7.2	10.4	10.6	10.0	18.4	17.8	16.2
	5	5.1	5.2	5.0	6.1	6.3	5.9	7.9	7.7	7.3	13.0	12.6	11.0
	10	4.9	5.0	4.8	5.4	5.6	4.8	5.9	5.9	5.6	9.1	8.6	6.7

TABLE 4.6. (*Continued*)

k	m	$\gamma_0 = .1$			$\gamma_0 = .2$			$\gamma_0 = .3$			$\gamma_0 = .5$		
		0	S	L	0	S	L	0	S	L	0	S	L
					(c) Nominal level $\alpha = 10\%$								
2	2	9.6	9.4	9.5	10.0	10.8	10.2	12.6	12.8	12.7	17.2	17.6	16.6
	3	9.7	9.7	9.6	9.9	10.3	9.9	11.0	11.2	10.5	13.4	14.0	13.3
	5	9.7	9.5	9.6	9.7	10.0	9.5	10.7	10.6	9.7	12.2	12.4	11.7
	10	9.6	9.6	9.7	9.5	9.9	9.5	10.4	10.1	9.5	11.2	11.4	10.9
3	2	9.8	9.5	9.4	11.2	11.7	11.6	14.1	14.4	13.8	19.8	20.3	20.1
	3	9.5	9.6	9.3	10.6	10.7	10.5	11.5	12.0	11.1	15.8	15.5	15.6
	5	9.6	9.7	9.3	10.5	10.4	9.9	10.5	10.7	9.5	13.3	13.3	12.8
	10	9.7	9.5	9.6	10.1	10.4	9.7	10.5	10.6	9.3	11.9	11.9	11.3
5	2	10.2	10.6	10.0	12.3	13.0	13.4	15.8	17.3	15.9	24.9	26.0	24.9
	3	9.9	10.1	9.6	11.1	11.5	11.1	13.4	13.5	12.9	18.9	18.9	18.0
	5	10.0	10.1	9.5	10.7	11.0	10.2	11.3	11.5	11.0	14.8	15.5	14.6
	10	9.9	9.8	9.3	10.3	10.4	9.6	10.3	10.4	9.8	12.6	12.8	12.3
10	2	11.0	11.2	10.8	15.8	16.0	16.1	21.1	22.5	21.4	31.3	31.9	30.2
	3	10.6	10.3	9.7	13.1	13.4	12.8	16.7	17.3	16.3	24.3	23.8	22.0
	5	10.4	10.1	9.4	11.5	11.4	10.7	13.7	13.6	12.6	19.4	18.6	16.2
	10	10.3	9.9	9.3	10.7	10.5	10.0	11.5	11.1	10.2	15.2	14.9	12.1

used in Tables 4.4–4.6. Clearly, even $\hat{D}_m$ with $m = 2$ is quite superior to single imputation (e.g., when $\alpha = 5\%$, $k = 10$ and $\gamma_0 = .3$, a 14.8% rejection rate for $\hat{D}_2$ versus a 45.3% rate for d_{*1}). The practical adequacy of $\hat{D}_m$ is supported by work reported in Weld (1987) with modest sample sizes and the data of Example 1.3.

TABLE 4.7. Large-sample level (in %) of d_{*1} with χ_k^2 reference distribution as a function of nominal level, α; number of components being tested, k; and fraction of missing information, γ_0.

k	$\gamma_0 =$	$\alpha = 1\%$				$\alpha = 5\%$				$\alpha = 10\%$			
		.1	.2	.3	.5	.1	.2	.3	.5	.1	.2	.3	.5
2		2.3	4.6	8.4	21.5	8.6	13.6	19.9	36.8	15.2	21.5	28.9	46.4
3		2.6	5.6	10.6	28.6	9.4	15.7	24.0	45.7	16.4	24.4	33.9	55.5
5		3.0	7.4	15.0	41.2	10.7	19.4	31.0	59.5	18.2	29.1	41.9	68.8
10		4.0	11.6	25.3	65.5	13.3	27.2	45.3	80.7	21.9	38.5	57.0	86.6

PROBLEMS

1. Suppose $C = C(X, Y_{inc}, I)$ is an interval estimate of Q that is free of R and that the sampling mechanism is unconfounded with R: $\Pr(I|X, Y, R) = \Pr(I|X, Y)$. Show that the fixed-response randomization-based coverage of C for Q is the same no matter what value R is fixed at:

$$\text{Prob}\{Q \in C | X, Y, R = R'\} = \text{Prob}\{Q \in C | X, Y, R = R''\}$$

for all R', R''. Furthermore, show that the random-response randomization-based coverage of C for Q is the same no matter what specification is made for the response mechanism; that is, show that

$$\text{Prob}\{Q \in C|X, Y\} \text{ takes the same value}$$
$$\text{for all specifications } \Pr(R|X, Y).$$

2. Comment on the possibility of using nonnormal reference distributions for randomization inference. In particular, let the estimand be $\overline{Y}$ and consider the following quotation:

> When I had the stones on the table around 1935, I had the whole population distribution which for weights was very skew, so that the randomization distribution of the means of samples of size five was still skew. I used Fisher's old method of approximating the [randomization distribution of the] means of five by using the first four moments of their distribution. (Personal communication from W. G. Cochran, July 13, 1979.)

3. Let $Z_i = \log(Y_i)$ where the objective is to draw inferences about $\overline{Y}$ from a simple random sample of size n; n is large and n/N is very small. Let $\bar{y}$ and $\bar{z}$ be the sample means of Y and Z, respectively, and s^2 and s_z^2 their sample variances. Consider the interval estimate of $\overline{Y}$

$$C(\bar{z}, s_z^2) = \exp(\bar{z} + s_z^2/2) \pm 1.96 \exp(\bar{z} + s_z^2/2) s_z (1 + s_z^2/2)^{1/2}/n^{1/2}.$$

(a) Show that if the Z_i are modeled as i.i.d. $N(\mu, \sigma^2)$ with a diffuse prior distribution on $\theta = (\mu, \sigma^2)$

$$E(\overline{Y}|Y_{inc}) = E\left[\exp(\mu + \sigma^2/2)|Y_{inc}\right] \to \exp(\bar{z} + s_z^2/2)$$

$$V(\overline{Y}|Y_{inc}) = V\left[\exp(\mu + \sigma^2/2)|Y_{inc}\right]$$
$$\to \exp(\bar{z} + s_z^2/2) V(\mu + \sigma^2/2|Y_{inc}),$$

where

$$V(\mu + \sigma^2/2 | Y_{inc}) = (s_z^2 + s_z^4/2)/n.$$

Hint: Try a first-term Taylor series expansion of $\exp(\mu + \sigma^2/2)$ about $(\bar{z} + s_z^2/2)$.

(b) Is $C(\bar{z}, s_z^2)$ an approximate 95% posterior interval for $\bar{Y}$?

4. Problem 3 continued. Assume the population values Z_i are i.i.d. $N(\mu_*, \sigma_*^2)$.

(a) Show that $C(\bar{z}, s_z^2)$ is an approximate 95% confidence interval for $\bar{Y}$:

$$E\left[\exp(\bar{z} + s_z^2/2) | Y\right] \to \exp(\mu_* + \sigma_*^2/2) = \bar{Y}$$

$$E\left[\exp(2\bar{z} + s_z^2)(s_z^2 + s_z^4/2)/n | Y\right] \to V\left[\exp(\bar{z} + s_z^2/2) | Y\right]$$

Hint: Now expand $\exp(\bar{z} + s_z^2/2)$ about $(\mu_* + \sigma_*^2/2)$.

(b) State why the standard 95% interval for $\bar{Y}$

$$C'(\bar{y}, s^2) = \bar{y} \pm 1.96 s/n^{1/2}$$

is a large-sample 95% confidence interval for $\bar{Y}$.

(c) Show that under the lognormal model

$$\bar{y} \to \exp(\mu_* + \sigma_*^2/2) = \bar{Y}$$

$$s^2 \to \exp(2\mu_* + 2\sigma_*^2)[1 - \exp(-\sigma_*^2)].$$

(d) Show that the ratio of the limiting widths of intervals $C(\bar{z}, s_z^2)$ and $C'(\bar{y}, s^2)$ is

$$\left[\frac{\sigma_*^2(1 + \sigma_*^2/2)}{\exp(\sigma_*^2) - 1}\right]^{1/2} = \left(\frac{1 + \sigma_*^2/2}{1 + \sigma_*^2/2 + \sigma_*^4/3! + \cdots}\right)^{1/2}.$$

(e) Which interval is better based on (d)?

5. Problem 4 continued. Is $C(\bar{z}, s_z^2)$ better than $C'(\bar{y}, s^2)$ in general, as when the Y_i are not lognormal? Hint: Consider an unrealistic but computationally trivial case; for example, suppose that $N/2$ of the Y_i are 1 and $N/2$ of the Y_i are 22,000, and show that the center of the standard interval, $\bar{y}$, tends toward $\bar{Y} = 11,000$, but, since as $n \to \infty$, $n/2$ of the Z_i will be 0 and $n/2$ will be 10, the center of the lognormal interval tends toward $\exp(5 + 12.5) = 4 \times 10^7$, thereby excluding the correct answer $\bar{Y}$ with probability 1 as $n \to \infty$.

6. Summarize the randomization-based evaluations of multiple-imputation procedures for scalar estimands given in Section 1 of this chapter.
7. Comment on the differences between the statistics $\hat{Q} = \hat{Q}(X, Y_{inc}, R_{inc}, I)$, where $Y_{inc} = (Y_{obs}, Y_{mis})$, and $\hat{Q}_{*l} = \hat{Q}_{*l}(X, Y_{inc,l}, R_{inc}, I)$ where $Y_{inc,l} = (Y_{obs}, Y_{mis,l})$ with $Y_{mis,l}$ the lth set of imputed values for Y_{mis}. From the randomization perspective, from what distribution does Y_{mis} arise? Hint: think fixed Y. From what distribution does $Y_{mis,l}$ arise? Hint: think posited model.
8. Define *randomization-valid*; which quantities are fixed?
9. When are infinite-m repeated-imputation inferences randomization-valid?
10. In (4.2.1) and (4.2.2), are $\hat{Q}$ and U complete-data statistics or can they be completed-data statistics?
11. Discuss the possibility of defining randomization-validity so that it holds only under the true model for nonresponse.
12. Describe a realistic case when the imputer's and data analyst's models are both correct or nearly so.
13. Summarize the definitions of proper and repeated multiple-imputation methods and present an example of each of the four types: proper and repeated, proper but not repeated, improper but repeated, and improper and not repeated.
14. Define proper imputation methods when B_∞ is not full rank.
15. Summarize Result 4.1 and combine it with Conclusion 4.1 to provide a practical summary.
16. Why do (4.3.1) and (4.3.2) follow from (4.2.3) and (4.2.4) under the assumptions?
17. Prove in the context of Example 4.1 that:

    (a) $\overline{Q}_\infty = \bar{y}_1$

    (b) $\overline{U}_\infty = s_1^2(n^{-1} - N^{-1})$

    (c) $E(\overline{U}_\infty | X, Y, I) \doteq U$ and $V(\overline{U}_\infty | X, Y, I) \ll B$

    (d) $V(B|X, Y) \ll E(U|X, Y)$

    (e) $E(B_\infty | X, Y, I) = B(n_1/n)$

18. How does Example 4.1 change if n_1 is treated as binomial (n, θ_R) where θ_R is the probability of response?
19. In Example 4.1, the length of the repeated-imputation confidence interval for Q is too short by what factor?

PROBLEMS

20. Repeat Problem 17 for the fully normal imputation method of Example 4.2 making appropriate changes in results.

21. Repeat Problem 20 for the Bayesian bootstrap of Example 4.3.

22. Repeat Problem 20 for the approximate Bayesian bootstrap of Example 4.4.

23. Repeat Problem 20 for MV multiple imputation of Example 4.5.

24. Repeat Problem 20 for the adjusted hot-deck multiple-imputation method of Rubin (1979a) and Herzog and Rubin (1983).

25. Present a more formally satisfying statement of Conclusion 4.1 and a proof of the revised statement.

26. Show that all the repeated-imputation methods of Examples 4.2–4.5 are asymptotically improper for $\{\bar{y}, s^2(n^{-1} - N^{-1})\}$ under the non-ignorable response mechanism used in the example of Section 1.6.

27. Relate Problems 3–5 to the discussion of the role of the complete-data statistics in determining whether the imputation method is proper.

28. Suppose the imputer's model for $\Pr(Y, X)$ is, in i.i.d. form,

$$\int \prod_{i=1}^{N} f(Y_i, X_i)|\theta) \Pr(\theta) \, d\theta,$$

and that the data analyst's model for $\Pr(Y, X)$ is identical *except* for the prior distribution on θ, which fixes some components of θ at null values.

(a) Show how this structure can accommodate a predictor used by the imputer but not the data analyst (e.g., an interaction term in a regression model used by the imputer but not the data analyst).

(b) In what senses are the imputations "safe" for the data analyst? For example, are significance levels conservative and are interval estimates conservatively calibrated assuming the data analyst's model is correct?

(c) Summarize any conclusions concerning the use of more versus less restrictive models for imputation, where restrictive refers to prior constraints that the data can address.

(d) Relate these conclusions to the idea that imputation models that generate data with more rather than less variability are safer.

29. Comment on the possibility of obtaining inferences from a multiply-imputed data set by means other than the repeated-imputation inferences derived in Chapter 3. For instance, Rubin (1977b) suggests the possibility of performing a weighted analysis: for m imputations, m pseudounits are created for each unit with missing data, where the m weights for the pseudounits add up to the total weight for that unit. What role does linearity of Q on Y play?

30. Can an "errors-in-variables" model be usefully employed to analyze a multiply-imputed data set by viewing the multiple imputations as flawed measurements of an underlying true (latent) value?

31. Suppose a simple random sample of size n of $Y_i = (Y_{i1}, Y_{i2})$ is taken from a bivariate normal population with $N/n \to \infty$; Y_{i1} is fully observed in the sample and Y_{i2} is partially missing due to ignorable nonresponse (i.e., missingness on Y_{i2} might depend on the value of Y_{i1}). The estimand Q is the regression coefficient of Y_{i1} on Y_{i2}. Do the multiple imputations of predictor variables Y_{i2} create an "errors-in-variables" bias in the regression coefficients? Justify your answer, both by general theory and by specific calculations.

32. Explicitly derive (4.5.9)–(4.5.11) from (4.2.3)–(4.2.9), (4.5.1), and (4.5.2); also derive the multivariate extension (4.5.16).

33. Justify the t reference distribution for inferences about scalar Q from the randomization perspective.

34. Derive (4.5.20) from (4.5.18).

35. Review the relationship between γ_0 and the EM algorithm.

36. For the BB, FN, MV, and ABB methods, show that
$$V(\bar{Q}_m | Y, R, I) = B_\infty / m = \left(n_1^{-1} - n^{-1} \right) E\left(s_0^2 | Y, R, I \right) / m.$$
Also, derive their values of $E(s_0^2 | Y, R, I)$.

37. Consider the following statement:

  If the reference distribution of $\bar{Q}_m$ were normal, the relative efficiency expression (4.5.20) would be entirely appropriate. The reference distribution for $\bar{Q}_m$ is not normal, however, but t on ν degrees of freedom, and thus, an adjustment should be made to the relative efficiency to reflect the increased inferential uncertainty associated with t distributions on finite degrees of freedom relative to normal distributions. The average second derivative of a t log posterior on ν degrees of freedom at the mode is $(\nu + 3)/(\nu + 1)$ times that of the limiting normal, and so a simple adjustment is to multiply the relative efficiency by $[(\nu + 1)/(\nu + 3)]^{1/2}$.

   (a) Discuss the issue of the need for such an adjustment from the randomization perspective. For example, if two estimates have the same variance but one has more degrees of freedom associated with its estimated variance, are the two estimates equally precise?
   (b) Derive the adjustment.
   (c) Would the adjusted or unadjusted efficiencies be more relevant to the creator of a multiply-imputed data set when deciding how large m should be and why?

PROBLEMS

38. Report the work in Rubin and Schenker (1986) concerning linear interpolation in $(m - 1)^{-1}$ for coverages.

39. Compare (4.6.4) with the Bayesian appraisal of the accuracy of the interval estimate using the t reference distribution. Hint: see Problem 12 in Chapter 3.

40. Assume Y_i is univariate and let

$$p_{*l} = \text{Prob}\{Q > 0 | X, Y_{obs}, Y_{mis,l}, R_{inc}\} = \Phi\left(\hat{Q}_{*l} / U_{*l}^{1/2}\right)$$

(a) Discuss the use of $\bar{p}_m = \sum_1^m p_{*l}/m$ to estimate $\text{Prob}\{Q > 0 | X, Y_{obs}, R_{inc}\}$; consider $m = 1$, $m = \infty$ and intermediate values and the actual level of the implied significance tests.

(b) Is $\bar{p}_m$ a reasonable estimate of $\bar{p}_\infty$? Comment on the effect of the prior distribution that restricts support to $(0, 1)$. Does a wiser prior lead to more realistic implied significance levels?

(c) Is there a better way to estimate $\bar{p}_\infty$? Hint: use $\bar{Q}_m, \bar{U}_m, B_m$.

(d) Assume all $U_{*l} = \bar{U}_\infty$; is there a better way to estimate $\bar{p}_\infty$ than $\bar{p}_m$ that is based only on the p_{*l}? Hint: consider the $\Phi^{-1}(p_{*l})$, their sample mean and variance, and part (c).

41. Show that $\text{Prob}\{D_m > F_{k,\nu}(\alpha) | X, Y; Q = Q_0\}$ asymptotically equals (4.7.8).

42. Describe efficient methods for evaluating the level and power of the test based on D_m.

43. Fit a response surface model to Table 4.4 and summarize conclusions.

44. Show that $\text{Prob}\{\tilde{D}_m > F_{k,k'\nu}(\alpha) | X, Y; Q = Q_0\}$ asymptotically equals (4.7.9).

45. Fit response surface models to Tables 4.5 and 4.6 and summarize conclusions.

46. Formulate and evaluate a procedure that for small m is nearly $\tilde{D}_m$ with the $F_{k,(k+1)\nu/2}$ reference distribution but for large m tends to D_m.

47. Compare the large-m powers of $\tilde{D}_m$ and D_m.

48. Compare the large-m powers of $\tilde{D}_m$ and $\hat{D}_m$, and evaluate how much is lost asymptotically by having to estimate r_m.

49. Formulate and evaluate a procedure better than $\hat{D}_m$ that uses only $\{d_{*l}\}$.

50. Derive (4.8.1), and justify using (4.8.2) to approximate (4.8.1) when the k fractions of missing information are not equal.

51. Summarize the conclusions of this chapter for the applied user of a multiply-imputed data set (e.g., a social scientist).

CHAPTER 5

Procedures with Ignorable Nonresponse

5.1. INTRODUCTION

The theory presented in Chapter 3 and 4 implies that if multiple imputations are drawn as repetitions from the posterior predictive distribution of the missing values under an appropriate model, or a reasonable approximation to it, then approximately valid inferences will result by properly combining the complete-data inferences. Although this theory is general, all the specific illustrations of it, and in particular, the illustrations of methods for creating multiple imputations in Chapter 4, were for the very simple case with a simple random sample of one outcome variable with no covariates and ignorable nonresponse, where the objective is to estimate the population mean using the standard inference. This case is unrealistically simple relative to most real survey situations. First, there are often many outcome variables (i.e., variables only observed for units included in the survey) and many covariates (i.e., variables fully observed for all units in the population), rather than one outcome variable and no covariates. Second, there are commonly many quantities to be estimated, such as tables of counts, correlations, and regression coefficients, rather than just means, and their estimates can be complicated functions of both X and Y. Third, the sampling mechanism, although commonly an unconfounded probability sampling mechanism, is usually not simple random sampling but depends explicitly on covariates used to define, for example, strata and clusters of sampling units. Fourth, nonresponse often creates complicated patterns of missing data in outcome variables. Fifth, nonresponse is rarely known to be ignorable.

INTRODUCTION

Multiple imputation was not designed for application in the simple case used for illustration in Chapter 4 where the theoretically correct answer is immediate, but rather for more complex cases where theoretically satisfactory answers are difficult to derive explicitly. In this chapter, we consider more general cases, but still restrict attention to ignorable sampling and response mechanisms; nonignorable nonresponse is considered in Chapter 6. It is natural to focus first on ignorable models because doing so leads to trying to adjust for all observed sources of bias between respondents and nonrespondents.

No Direct Evidence to Contradict Ignorable Nonresponse

An important feature of the assumption of ignorable nonresponse is that generally there will be no direct evidence in the data to contradict it. Consider a group of respondents and nonrespondents with identical values of the observed variables X, and for simplicity, suppose that Y is univariate. The assumption of ignorable nonresponse means that the unobserved distribution of Y for the nonrespondents is only randomly different from the observed distribution of Y for the respondents. Since no Y values are observed for nonrespondents, without external information there will be no way to judge whether the nonrespondents' missing values are systematically different from the respondents' observed values. As an example of such external information, suppose that we expected the population distribution of values at the common observed value of X to be approximately symmetric; if the respondents' values are observed to be quite skew, then the observed values of Y coupled with this external knowledge constitute evidence of nonignorable nonresponse. But without some such external knowledge, the assumption of ignorable nonresponse is perfectly plausible since the population distribution of values of Y at this value of X might be skew just like the observed values of Y for respondents.

Adjust for All Observed Differences and Assume Unobserved Residual Differences Are Random

Imputation methods that assume ignorable nonresponse thus play an obvious central role: adjust for all observed differences between respondents and nonrespondents and assume that unobserved differences are random. If systematic unobserved differences are plausible, then these can be investigated as deviations from the ignorable imputed values. The artificial example in Section 1.6 did this by asserting that each nonignorable missing value was 20% higher than the associated ignorable missing value.

Of course, explicit nonignorable models, such as ones that capitalize on normality assumptions, can be built, as discussed and illustrated in Chapter 6.

Univariate Y_i and Many Respondents at Each Distinct Value of X_i That Occurs Among Nonrespondents

Possibly the simplest case with ignorable nonresponse, beyond the rather artificial one of a simple random sample of univariate Y_i used to illustrate ideas in Chapter 4, arises when Y_i is univariate and for each nonrespondent there are many respondents with the same value of X_i. With such a situation, it may be reasonable to treat all the units with a common value of X_i as a separate group and apply one of the multiple-imputation methods used in Examples 4.2–4.5 applicable to a simple random sample with ignorable nonresponse and univariate Y_i.

Formally, if there are K distinct values of X_i, we are fitting independent models at each of the K values of X_i; using the usual i.i.d. formulation, we are allocating to each distinct value of X_i an independent parameter (possibly a vector) governing the distribution of Y_i. This assumption of independence across distinct X_i is most appropriate in cases with many respondents at each observed value of X_i from which to estimate the distribution of Y_i given X_i.

The More Common Situation, Even with Univariate Y_i

More commonly, there will be some X_i values occurring among the nonrespondents that will occur among only few if any respondents; this is especially true with multivariate X_i. For example, if X_i has 20 binary components, there are over a million possible values of X_i; and it may be quite likely that one particular value occurs for a nonrespondent but not for any respondent, even if respondents greatly outnumber and only randomly differ from nonrespondents.

There are basically two standard approaches for imputing data in this situation. The first approach relies on implicit models that define collections of respondents who are "close" to each nonrespondent and thereby returns to imputation methods for univariate Y_i; in simple random samples, this approach dominates current survey practice. The second approach builds explicit statistical models for the conditional distribution of Y_i given X_i, such as a normal linear regression model, and creates imputations based on the model. A key issue with implicit modeling methods is defining "close," since with high-dimensional X_i and a modest number of respondents, the closest available on X_i according to one metric (i.e., definition of closeness)

INTRODUCTION

may be very far away with respect to another definition, and in particular with respect to the distribution of Y_i given X_i. That is, consider a nonrespondent with $X_i = x'$; even if a respondent with $X_i = x''$ is the closest available respondent with respect to a certain metric, $\Pr(Y_i | X_i = x')$ may be very different from $\Pr(Y_i | X_i = x'')$, especially so when the distributions of X are different for respondents and nonrespondents and the dimensionality of X is large.

The key practical issue with the explicit modeling approach is defining the global model for $\Pr(Y_i|X_i)$ to be accurate locally. In particular, suppose $\Pr(Y_i|X_i)$ specifies that univariate Y_i has a linear regression on univariate X_i, but in fact, $\log(Y_i)$ has a linear regression on X_i; the linear approximation can be a decent global approximation, accounting for a large proportion of the variance of Y due to X, but still can create less accurate imputations than an implicit model that matches on X. Nonparametric regression techniques can often be viewed as falling between matching methods and parametric regression approaches.

A Popular Implicit Model—The Census Bureau's Hot-Deck

A quite popular method for handling nonresponse makes all X variables categorical and tries to find, for each nonrespondent, exactly matching respondents with respect to the categorical X; the U.S. Census Bureau's hot-deck for the Current Population Survey uses such a method. If a matching respondent is found for a nonrespondent, then the respondent donates its values to the nonrespondent. If more than one respondent matches the nonrespondent, then, depending on the particular implementation, either the first or a randomly chosen respondent is the donor. (Census files are often ordered by potentially relevant characteristics, such as geographical location.) If no matching respondents are found for a nonrespondent, some categories of X are made coarser, or some components of X are dropped altogether (the coarsest a variable can be—one category) according to rather complicated rules, and the procedure is tried again. The process continues with coarser and coarser X until a matching respondent is found. For example, if the first component of X is state in the United States, the first pass for a nonrespondent may use a nine-region classification, the second pass a four-region classification, and all subsequent passes may drop the state variable altogether (i.e., use one region—the entire United States). Different rules can thus be applied to different nonrespondents depending on how easy it is to find matching respondents. All nonrespondents, however, will eventually receive a matching respondent. Recent descriptions of hot-deck procedures appear in Ford (1983) and Sande (1983).

A simple multiple-imputation version would modify the rules to require a minimum of $m \geq 2$ rather than 1 matching respondent, where the first m or a random m would be chosen as donors if there were more than m matching respondents. With univariate Y_i, the methods in Examples 4.2–4.5 could be directly applied to the collection of potential donors to create multiple imputations.

Metric-Matching Hot-Deck Methods

Metric-matching hot-deck methods define a measure of distance between each nonrespondent and each respondent, such as $(x' - x'')S^{-1}(x' - x'')'$ where x' and x'' refer to specific nonrespondent and respondent values of X_i and S is the covariance matrix of X_i in the respondent sample. With multiple imputation, the matching respondents for a nonrespondent could be the m closest respondents, or those m supplemented with all those respondents less than some fixed distance D_0 away from the nonrespondent, or possibly a random choice of m from those that are less than D_0 away. Thus, each nonrespondent will have available at least m matching nonrespondents to which the methods in Examples 4.2–4.5 can be applied when Y_i is univariate. Statistics Canada uses a version of metric matching for some of its single imputations (Sande, 1983).

Much of the statistical literature on matching methods appears in the context of observational studies for causal effects. In this context, for each exposed unit (e.g., smoker) a matching nonexposed control unit (nonsmoker) is sought (see, e.g., Cochran, 1968; Cochran and Rubin, 1973; Rubin, 1973, 1979b). In fact, the Census Bureau's hot-deck is quite similar to the "variable-caliper" matching method described by Althauser and Rubin (1971). Some of these methods are full rank so that the distance between x' and x'' can only be zero when $x' = x''$. Other methods, such as discriminant matching (Rubin, 1976b, 1980c) and propensity score matching (Rosenbaum and Rubin, 1983, 1985) define distance by creating a scalar summary of X_i that predicts group membership, that is, *treated* versus *control* in the context of observational studies and *nonrespondent* versus *respondent* in the context of nonresponse.

One important difference between the survey and observational study contexts is that at the time of the matching, the outcome variable Y is observed for survey respondents but generally not yet observed for any units in the observational study. Consequently, the matching in the nonresponse context can take explicit advantage of estimated relationships between Y and X in the respondent group. For example, with univariate Y_i, a regression of Y_i on X_i can be computed among respondents to define a scalar function of X_i that is best correlated with Y_i in this group; this scalar

INTRODUCTION

variable can then be computed for both respondents and nonrespondents and used to define acceptable matches. Of course, since a model is being used to estimate the relationship between Y_i and X_i among respondents, it becomes natural to consider using the model for the imputation itself rather than just for defining matches.

Least-Squares Regression

A common method for imputing missing values is via least-squares regression (e.g., Afifi and Elashoff, 1969). Regress univariate Y_i on X_i using the respondent data to obtain a prediction equation of the form $\hat{Y}_i = a + bX_i$, and then impute the missing Y_i for nonrespondents using this equation and the observed values of X_i. One obvious problem with such a best-prediction method, already indicated in Example 1.5, is that the imputed values will all fall on the estimated regression line and so will lead to obvious biases in estimands that involve the residual variance of Y_i given X_i for nonrespondents, which is estimated to be zero. Simple methods that attend to this problem draw residuals e_i with mean zero to add to $\hat{Y}_i$ before imputation, where the e_i can be drawn, for example, from (a) $N(0, s_e^2)$, where s_e^2 is the estimated residual variance of Y_i given X_i among respondents or (b) the actual empirical residuals among the respondents. In a multiple-imputation context, several imputed values would be created for each missing Y_i, where ideally uncertainty due to the estimation of the regression itself would be reflected across the imputations. Details of such a method are given in Section 5.3.

Outline of Chapter

In order to define precisely the steps to be kept in mind when creating multiple imputations, we need to consider how to create imputations under an explicit Bayesian model. As concluded in Section 4.4, this is theoretically and practically the most direct way to create imputations that lead to valid inference from both the Bayesian and randomization perspectives. The steps for creating imputations under an explicit Bayesian model are given in the relatively theoretical Section 5.2, followed by examples of explicit imputation models for univariate Y_i with covariates in Section 5.3. Methods for univariate Y_i with covariates can be directly applied to a much larger class of problems, those with monotone patterns of missing data in multivariate Y_i, and this generalization is discussed in Section 5.4 and illustrated using an example with bivariate Y_i and covariates in Section 5.5. Section 5.6 considers extensions to nonmonotone patterns of missing data.

5.2. CREATING IMPUTED VALUES UNDER AN EXPLICIT MODEL

Three formal tasks can be defined that are needed to create imputed values that simulate the posterior distribution of Y_{mis} under an explicit Bayesian model: the modeling task, the estimation task, and the imputation task. These were illustrated in the simple case of Examples 4.2–4.5 without formal definition. Here, we provide definitions and derive properties of the tasks in the general case with ignorable nonresponse.

The modeling task chooses a specific model for the data. The estimation task formulates the posterior distribution of the parameters of that model so that a random draw from it can be made. The imputation task takes one random draw from the posterior distribution of Y_{mis} by first drawing a parameter from the posterior distribution obtained in the estimation task, and then drawing Y_{mis} from its conditional posterior distribution given the drawn value of the parameter; the imputation task is repeated m times to create m imputations for each missing value under the chosen model.

The Modeling Task

The repeated multiple imputations for the missing values, Y_{mis}, represent m draws from the posterior distribution of Y_{mis} under the chosen model. Since sampling and nonresponse mechanisms are being assumed ignorable, by Result 2.3 the posterior distribution of Y_{mis} is its conditional distribution given the observed values X and Y_{obs}, $\Pr(Y_{mis}|X, Y_{obs})$. The posterior distribution of Y_{mis} follows from the specification for $\Pr(X, Y)$: $\Pr(Y_{mis}|X, Y_{obs}) = \Pr(X, Y)/\int \Pr(X, Y)\, dY_{mis}$. The modeling task provides a specification for $\Pr(X, Y)$.

Throughout, we assume $\Pr(X, Y)$ is modeled in i.i.d. form:

$$\Pr(X, Y) = \int \Pr(X, Y|\theta)\Pr(\theta)\, d\theta = \int \left[\prod_{i=1}^{N} f_{XY}(X_i, Y_i|\theta)\right] \Pr(\theta)\, d\theta.$$

(5.2.1)

From (5.2.1) we see that in a formal sense, the modeling task is simply a standard Bayesian specification of a model appropriate for a multivariate data set; for example, $f_{XY}(\cdot|\cdot)$ could be a $(p + q)$-variate normal for continuous data, or a $(p + q)$-dimensional log-linear model for discrete data, where p and q are the number of components in Y_i and X_i, respectively. The issues that arise with incomplete data are basically the same as arise with complete data, although in the context of imputation, there may be somewhat more emphasis on formulating models that will give accurate predictions of missing values. With many missing values, extra

CREATING IMPUTED VALUES UNDER AN EXPLICIT MODEL

care may be needed to formulate useful models, because there may not be enough observed values to appeal to standard arguments. For example, if the first two components of Y are modeled as bivariate normal, but these two components are never simultaneously observed, standard procedures for estimating the correlation between the two components are not immediately appropriate.

It is often convenient to write

$$f_{XY}(X_i, Y_i|\theta) = f_{Y|X}(Y_i|X_i, \theta_{Y|X})f_X(X_i|\theta_X), \qquad (5.2.2)$$

where $\theta_{Y|X}$ and θ_X are functions of θ. For example, $f_{Y|X}(\cdot|\cdot)$ with continuous Y_i could be a general normal linear model for the p-variate outcome given the q-variate predictor, or with discrete Y_i it could be a multivariate logistic regression model. Specific examples appeared in Section 2.5 and Problems 11–14 of Chapter 2, and more examples appear later.

Factoring the joint distribution of (X_i, Y_i) as in (5.2.2) can simplify model building efforts whenever $\theta_{Y|X}$ and θ_X are chosen to be *a priori* independent. Specifically, because X_i is always fully observed, when $\theta_{Y|X}$ and θ_X are modeled as *a priori* independent, they are *a posteriori* independent, and moreover, the posterior distribution of $\theta_{Y|X}$ does not involve the specification for $f_X(\cdot|\cdot)$; this result follows from general results in Rubin (1974) concerning estimation in i.i.d. models with missing data, and is proved directly in Results 5.1 and 5.2. The practical consequence is that when $\theta_{Y|X}$ is *a priori* independent of θ_X, no specification is needed for either $f_X(X_i|\theta_X)$ or the prior distribution of θ_X.

Notwithstanding this simplification, there are occasions, however, when it can be useful to have the parameters $\theta_{Y|X}$ and θ_X *a priori* dependent. For instance, if it is thought that the correlational structure of (X, Y) resembles an intraclass correlation matrix (i.e., all off-diagonal elements equal), the correlations among X variables, which are always observed, help to estimate the correlations among the Y variables and between the Y variables and the X variables; in such a case, $\theta_{Y|X}$ and θ_X should be modeled as *a priori* dependent.

The Imputation Task

The objective here is to draw Y_{mis} from its posterior distribution under the model (5.2.1). This posterior distribution can be written as

$$\Pr(Y_{mis}|X, Y_{obs}) = \int \Pr(Y_{mis}|X, Y_{obs}, \theta)\Pr(\theta|X, Y_{obs})\,d\theta.$$

Our plan is first to draw a value of θ from its posterior distribution, $\Pr(\theta|X, Y_{obs})$, say θ^*, and then draw a value of Y_{mis} from its conditional posterior distribution given the drawn value of θ, $\Pr(Y_{mis}|X, Y_{obs}, \theta = \theta^*)$. Repeating this process m times creates m draws from the joint posterior distribution of (Y_{mis}, θ). Ignoring the drawn values of θ gives m draws from the marginal posterior distribution of Y_{mis}. The reason for drawing values of both θ and Y_{mis} is that under the i.i.d. model (5.2.1), the task of drawing Y_{mis} given θ is relatively simple.

This fact is particularly clear in the simple case of univariate Y_i. Letting $\theta^*_{Y|X}$ be a value of $\theta_{Y|X}$ drawn from its posterior distribution, each missing Y_i is imputed independently according to $f_{Y|X}(Y_i|X_i, \theta^*_{Y|X})$. For example, if $f_{Y|X}(\cdot|\cdot)$ is the normal linear regression model with $\theta_{Y|X} = (\beta, \sigma^2)$, then the missing Y_i are drawn as independent normals with means $X_i\beta^*$ and variances σ^{*2} where the asterisk superscripts refer to the drawn values of the parameters. A new set of imputations is created by drawing a new value of $\theta_{Y|X}$; when m values of $\theta_{Y|X}$ are drawn, m imputations are made for each missing value.

With multivariate Y_i, some extra notation is helpful for stating exactly what is involved in the imputation task. Let $mis(i) = \{j|I_{ij} = 1$ and $R_{ij} = 0\}$ = the indices of missing Y values for unit i, and let $obs(i) = \{j|I_{ij} = 1$ and $R_{ij} = 1\}$ = the indices of observed Y values for unit i, $mis(i) \cup obs(i) = inc(i)$ and $inc(i) \cup exc(i) = \{1, \ldots, p\}$ in an obvious notation. For notational simplicity, let $Y_{i, mis(i)} = Y_{i, mis}$ and $Y_{i, obs(i)} = Y_{i, obs}$.

We are now ready to describe the imputation task. Let $\theta^*_{Y|X}$ be a value of $\theta_{Y|X}$ drawn from its posterior distribution. For each unit with missing data (i.e., for each $i \in ms = \{i|I_{ij} = 1$ and $R_{ij} = 0$ for some $j\}$), find the conditional distribution of $Y_{i, mis}$ given $X_i, Y_{i, obs}$ and $\theta_{Y|X} = \theta^*_{Y|X}$,

$$\Pr(Y_{i, mis}|X_i, Y_{i, obs}, \theta^*_{Y|X}) = \frac{\int f_{Y|X}(Y_i|X_i, \theta^*_{Y|X}) \, dY_{i, exc}}{\int\int f_{Y|X}(Y_i|X_i, \theta^*_{Y|X}) \, dY_{i, exc} \, dY_{i, mis}}, \quad (5.2.3)$$

and draw one value of $Y_{i, mis}$ to impute; the units are imputed independently. When the patterns of missing data vary across the units, the distributions that must be found [i.e., the results of the integrations in (5.2.3)] will also vary. A new set of imputations is created by drawing a new value of $\theta_{Y|X}$. Result 5.1 justifies this procedure of drawing the $Y_{i, mis}$ independently across units.

Result 5.1. The Imputation Task with Ignorable Nonresponse

Given θ, the $Y_{i, mis}$ are *a posteriori* independent with distribution depending on θ only through $\theta_{Y|X}$:

$$\Pr(Y_{mis}|X, Y_{obs}, \theta) = \prod_{ms} \Pr(Y_{i, mis}|X_i, Y_{i, obs}, \theta_{Y|X}). \quad (5.2.4)$$

CREATING IMPUTED VALUES UNDER AN EXPLICIT MODEL

The proof begins by using Bayes's theorem and the fact that $Y_{nob} = (Y_{exc}, Y_{mis})$, to obtain

$$\Pr(Y_{mis}|X, Y_{obs}, \theta) = \frac{\int \Pr(X, Y|\theta)\, dY_{exc}}{\int \Pr(X, Y|\theta)\, dY_{nob}},$$

which by (5.2.1) and (5.2.2) yields

$$\Pr(Y_{mis}|X, Y_{obs}, \theta) = \frac{\prod_{i=1}^{N} f_X(X_i|\theta_X) \int \left[\prod_{i=1}^{N} f_{Y|X}(Y_i|X_i, \theta_{Y|X})\right] dY_{exc}}{\prod_{i=1}^{N} f_X(X_i|\theta_X) \int \left[\prod_{i=1}^{N} f_{Y|X}(Y_i|X_i, \theta_{Y|X})\right] dY_{nob}}. \tag{5.2.5}$$

Obviously, the factors involving θ_X cancel, and the remaining expression involves only $\theta_{Y|X}$. Thus

$$\Pr(Y_{mis}|X, Y_{obs}, \theta) = \Pr(Y_{mis}|X, Y_{obs}, \theta_{Y|X}). \tag{5.2.6}$$

For notational convenience, let $dY_{exc} = \prod_{i=1}^{N} dY_{i, exc}$ and $dY_{nob} = \prod_{i=1}^{N} dY_{i, exc}\, dY_{i, mis}$, where if $exc(i)$ is null, $\int f_{Y|X}(Y_i|X_i, \theta_{Y|X})\, dY_{i, exc} = f_{Y|X}(Y_i|X_i, \theta_{Y|X})$ and analogously, if $mis(i)$ is null, $\int f_{Y|X}(Y_i|X_i, \theta_{Y|X})\, dY_{i, mis} = f_{Y|X}(Y_i|X_i, \theta_{Y|X})$. Then, from (5.2.5) and (5.2.6)

$$\Pr(Y_{mis}|X, Y_{obs}, \theta) = \prod_{i=1}^{N} \frac{\int f_{Y|X}(Y_i|X_i, \theta_{Y|X})\, dY_{i, exc}}{\int\int f_{Y|X}(Y_i|X_i, \theta_{Y|X})\, dY_{i, exc}\, dY_{i, mis}}.$$

Consider the ith factor in this product: if $mis(i)$ is null, that is, if $i \notin ms$, then the numerator and denominator are equal. Hence

$$\Pr(Y_{mis}|X, Y_{obs}, \theta) = \prod_{ms} \frac{\int f_{Y|X}(Y_i|X_i, \theta_{Y|X})\, dY_{i, exc}}{\int\int f_{Y|X}(Y_i|X_i, \theta_{Y|X})\, dY_{i, exc}\, dY_{i, mis}}, \tag{5.2.7}$$

which is equivalent to (5.2.4).

The Estimation Task

Result 5.1 shows that the only function of θ that is needed for the imputation task is $\theta_{Y|X}$, the parameter of $f_{Y|X}(Y_i|X_i, \theta_{Y|X})$ specified in the modeling task. The objective of the estimation task is to compute the posterior distribution of $\theta_{Y|X}$ so that a random draw of $\theta_{Y|X}$ can be made. In special cases, such as with standard normal linear models, it is easy to draw $\theta_{Y|X}$ from its posterior distribution. Generally, however, calculating the posterior distribution of $\theta_{Y|X}$ can be not only analytically intractable but also computationally demanding. Consequently, we often must be satisfied with approximate posterior distributions from which we can easily draw and consider using special Monte Carlo techniques to make adjustments for having drawn from a less than adequate approximation. Even though posterior distributions can still be intractable, substantial simplification can occur when $\theta_{Y|X}$ and θ_X are *a priori* independent.

Result 5.2. The Estimation Task with Ignorable Nonresponse When $\theta_{Y|X}$ and θ_X Are *a Priori* Independent

Suppose $\theta_{Y|X}$ and θ_X are *a priori* independent

$$\Pr(\theta) = \Pr(\theta_{Y|X})\Pr(\theta_X). \tag{5.2.8}$$

Then they are *a posteriori* independent; moreover, the posterior distribution of $\theta_{Y|X}$ involves only (a) the specifications $f_{Y|X}(\cdot|\cdot)$ and $\Pr(\theta_{Y|X})$, and (b) data from units with some Y_{ij} observed.

The proof is direct but requires several steps. From (5.2.1), (5.2.2), and (5.2.8),

$$\Pr(\theta|X, Y_{obs})$$

$$= \frac{\int \left[\prod_{i=1}^{N} f_{Y|X}(Y_i|X_i, \theta_{Y|X}) f_X(X_i|\theta_X)\right] \Pr(\theta_{Y|X})\Pr(\theta_X)\, dY_{nob}}{\iiint \left[\prod_{i=1}^{N} f_{Y|X}(Y_i|X_i, \theta_{Y|X}) f_X(X_i|\theta_X)\right] \Pr(\theta_{Y|X})\Pr(\theta_X)\, dY_{nob}\, d\theta_{Y|X}\, d\theta_X}$$

or

$$\Pr(\theta|X, Y_{obs}) = \frac{\prod_{i=1}^{N} f_X(X_i|\theta_X)\Pr(\theta_X)}{\int \left[\prod_{i=1}^{N} f_X(X_i|\theta_X)\right]\Pr(\theta_X)\, d\theta_X}$$

$$\times \frac{\int \left[\prod_{i=1}^{N} f_{Y|X}(Y_i|X_i, \theta_{Y|X})\right]\Pr(\theta_{Y|X})\, dY_{nob}}{\iint \left[\prod_{i=1}^{N} f_{Y|X}(Y_i|X_i, \theta_{Y|X})\Pr(\theta_{Y|X})\right] dY_{nob}\, d\theta_{Y|X}},$$

where the first factor does not involve $\theta_{Y|X}$ and thus is $\Pr(\theta_X|X, Y_{obs})$, and the second factor does not involve θ_X and thus is $\Pr(\theta_{Y|X}|X, Y_{obs})$. Hence, $\theta_{Y|X}$ and θ_X are *a posteriori* independent,

$$\Pr(\theta|X, Y_{obs}) = \Pr(\theta_{Y|X}|X, Y_{obs})\Pr(\theta_X|X, Y_{obs}), \tag{5.2.9}$$

CREATING IMPUTED VALUES UNDER AN EXPLICIT MODEL

with

$$\Pr(\theta_{Y|X}|X, Y_{obs}) = \frac{\left[\prod_{i=1}^{N} \int f_{Y|X}(Y_i|X_i, \theta_{Y|X})\, dY_{i,nob}\right] \Pr(\theta_{Y|X})}{\int \left[\prod_{i=1}^{N} f_{Y|X}(Y_i|X_i, \theta_{Y|X})\, dY_{i,nob}\right] \Pr(\theta_{Y|X})\, d\theta_{Y|X}},$$

(5.2.10)

which involves only the specifications $f_{Y|X}(\cdot|\cdot)$ and $\Pr(\theta_{Y|X})$. The product $\prod_{i=1}^{N}$ in (5.2.10) can be replaced by $\prod_{i \in ob}$ where $ob = \{i | I_{ij} R_{ij} = 1$ for some $j\}$ = the set of units with some Y_{ij} observed, because for the other units the factors are one, whence $\Pr(\theta_{Y|X}|X, Y_{obs})$ involves only units with some Y_{ij} observed:

$$\Pr(\theta_{Y|X}|X, Y_{obs}) = \frac{\left[\prod_{ob} \int f_{Y|X}(Y_i, \theta_{Y|X})\, dY_{i,nob}\right] \Pr(\theta_{Y|X})}{\int \left[\prod_{ob} \int f_{Y|X}(Y_i, \theta_{Y|X})\, dY_{i,nob}\right] \Pr(\theta_{Y|X})\, d\theta_{Y|X}}.$$

(5.2.11)

Result 5.3. *The Estimation Task with Ignorable Nonresponse, $\theta_{Y|X}$ and θ_X a Priori Independent, and Univariate Y_i*

If $\theta_{Y|X}$ and θ_X are *a priori* independent and Y_i is univariate so that the respondents have Y_i observed and the nonrespondents are missing Y_i, the posterior distribution of $\theta_{Y|X}$ involves only the respondents.

Result 5.3 is an immediate consequence of Result 5.2, and yields (5.2.12):

$$\Pr(\theta_{Y|X}|X, Y_{obs}) = \frac{\left[\prod_{obs} f_{Y|X}(Y_i|X_i, \theta_{Y|X})\right] \Pr(\theta_{Y|X})}{\int \left[\prod_{obs} f_{Y|X}(Y_i|X_i, \theta_{Y|X})\right] \Pr(\theta_{Y|X})\, d\theta_{Y|X}}. \quad (5.2.12)$$

It is stated separately because the case of univariate Y_i is basic to many practical imputation methods. From Result 5.1, with univariate Y_i the posterior distribution of Y_{mis} given θ is simply

$$\Pr(Y_{mis}|X, Y_{obs}, \theta) = \prod_{mis} f_{Y|X}(Y_i|X_i, \theta_{Y|X}) \quad (5.2.13)$$

A Simplified Notation

For simplicity, henceforth in this chapter we assume that $\theta_{Y|X}$ and θ_X are *a priori* independent, and thus we need only be concerned with the specifications $f_{Y|X}(Y_i|X_i, \theta_{Y|X})$ and $\Pr(\theta_{Y|X})$. For notational simplicity, we drop the subscript $Y|X$ and write $f(Y_i|X_i, \theta)$ and $\Pr(\theta)$ for these specifications:

$$\Pr(Y|X) = \int \Pr(Y|X, \theta)\Pr(\theta)\, d\theta \qquad (5.2.14)$$

where

$$\Pr(Y|X, \theta) = \prod_{i=1}^{N} f(Y_i|X_i, \theta). \qquad (5.2.15)$$

5.3. SOME EXPLICIT IMPUTATION MODELS WITH UNIVARIATE Y_i AND COVARIATES

In order to help fix ideas presented in the rather theoretical Section 5.2, we consider univariate Y_i and several examples of multiple-imputation procedures based on explicit models. The case of univariate Y_i with covariates is actually far more general than it might appear because in many common cases, outlined in Section 5.4 and illustrated in Section 5.5, multiple imputations for multivariate Y_i can be created by repeated application of methods designed for univariate Y_i. We use the simplified notation in (5.2.14) and (5.2.15) appropriate when the parameter for the distribution of Y given X is independent of the parameter for the distribution of X.

Example 5.1. Normal Linear Regression Model with Univariate Y_i

Perhaps the most common method of predicting univariate Y_i from a collection of predictors X_i is the normal linear regression model. Here,

$$Y_i \sim N(X_i\beta, \sigma^2)$$

is the specification for $f(Y_i|X_i, \theta)$, $\theta = (\beta, \log \sigma)$, β a vector of q components and σ a scalar. We complete the modeling task by assuming the conventional improper prior for θ, $\Pr(\theta) \propto$ const; to avoid irrelevant complexities, we also assume $n_1 > q$, where n_1 is the number of respondents.

We are now ready to describe the estimation task. First, note from Result 5.3 that the posterior distribution of θ involves only the units with Y_i observed. Standard Bayesian calculations with the normal linear model, which are described, for example, in Box and Tiao (1973) and here in

SOME EXPLICIT IMPUTATION MODELS WITH UNIVARIATE Y_i

Problem 14 in Chapter 2, show that *a posteriori*, σ^2 is $\hat{\sigma}_1^2(n_1 - q)$ divided by a $\chi^2_{n_1-q}$ random variable, and β given σ^2 is normal with mean $\hat{\beta}_1$ and variance–covariance matrix $\sigma^2 V$, where, in terms of the usual least-squares statistics based on the n_1 vectors (Y_i, X_i), $i \in obs$,

$$\hat{\sigma}_1^2 = \sum_{obs} (Y_i - X_i \hat{\beta}_1)^2 / (n_1 - q)$$

$$\hat{\beta}_1 = V \left[\sum_{obs} X_i' Y_i \right],$$

where

$$V = \left[\sum_{obs} X_i' X_i \right]^{-1}.$$

Since we have described the posterior distribution of θ in terms of standard distributions from which we can easily draw, the estimation task is complete.

We are finally ready to describe the imputation task for this model:

1. Draw a $\chi^2_{n_1-q}$ random variable, say g, and let

$$\sigma_*^2 = \hat{\sigma}_1^2 (n_1 - q)/g.$$

2. Draw q independent $N(0,1)$ variates to create a q-component vector Z and let

$$\beta_* = \hat{\beta}_1 + \sigma_* [V]^{1/2} Z,$$

where $[V]^{1/2}$ is a square root of V such as the triangular square root obtained by Cholesky factorization.

3. Draw the n_0 values of Y_{mis} as

$$Y_{i*} = X_i \beta_* + z_i \sigma_*,$$

where the n_0 normal deviates z_i are drawn independently.

A new imputed value for Y_{mis} is initiated by drawing a new value of the parameter σ_*^2. Thus, if m repeated imputations are desired, these three steps are repeated m independent times.

Example 5.2. Adding a Hot-Deck Component to the Normal Linear Regression Imputation Model

In Example 4.5 a hot-deck component was added to the fully normal imputation method of Example 4.2. Similarly a hot-deck component can be added to the linear regression imputation method of Example 5.1. Simply replace the drawing of the n_0 $N(0,1)$ deviates called for in step (3) of Example 5.1 with the drawing of n_0 values with replacement from the set of n_1 observed residuals standardized to have variance 1: $\{(Y_i - \hat{\beta}_1 X_i) \times (1 - q/n_1)^{-1/2}/\hat{\sigma}_1 | i \in obs\}$. Such a modification may help to preserve subtle deviations from normality of the residuals. In order to help preserve subtle deviations from linearity, it may be wise to exclude some of the n_1 residuals as being impossible, for example, by using a hot-deck method to define acceptable matches on the basis of X.

An even more extreme departure from the normal linear regression imputation technique in Example 5.1 is to replace step (3) with the following step used in Rubin (1986) as an illustration:

(a) Calculate the n_0 predicted values in Y_{mis} as

$$Y_{i*} = X_i \beta_* \qquad i \in mis.$$

(b) For each Y_{i*} $i \in mis$, find the respondent whose Y_i ($i \in obs$) is closest to Y_{i*}, and impute this value for Y_i.

This method creates between-imputation variability since it uses the normal linear regression steps (1) and (2) to draw β_*, and uses a linear model to guide the choice of values to impute, yet imputes only values already observed in Y_{obs}.

Extending the Normal Linear Regression Model

Of course, there is no reason why the X variables used in the regression models of Examples 5.1 or 5.2 should be identical to the X variables as initially recorded; functions of X variables may be needed to make the regression of Y_i on X_i more nearly a linear one with constant residual variance. For example, if the first component of X_i measures size in terms of number of employees, it may be more sensible to assume Y_i is linearly related to log(size) than to size itself. The exact same methods can be used no matter what transformations of X_i are employed.

In some cases, it may be unreasonable to assume normality for the regression of Y_i on X_i. One alternative class of models arises from assuming

SOME EXPLICIT IMPUTATION MODELS WITH UNIVARIATE Y_i

that some fixed transformation of Y_i, such as $\log(Y_i)$ has a normal linear regression on X_i, or for that matter on functions of X_i. Or the functions to be regressed on X_i can be functions of both X_i and Y_i; for example, with scalar X_i, Y_i/X_i might be modeled as having a normal linear regression on X_i with variance proportional to X_i (see Problems 11–14 in Chapter 2). Any such fixed transformation of Y_i creates no problem when generating multiple imputations of Y_{mis}. Simply derive the posterior distribution of parameters in terms of the transformed Y_i, say $\log(Y_i)$, and generate n_0 imputed values for the missing $\log(Y_i)$; before imputing values in Y_{mis}, invert the transformation, for example, take $\exp(\log Y_i)$, to create n_0 imputed values for Y_{mis}. When Y_i is dichotomous or takes on only a few values, a logistic regression model is likely to be more appropriate than a linear regression model.

Example 5.3. A Logistic Regression Imputation Model for Dichotomous Y_i

Suppose that Y_i is dichotomous (0 – 1) and that

$$f(Y_i|X_i, \theta) = \text{logit}^{-1}(X_i\theta)^{Y_i}\left[1 - \text{logit}^{-1}(X_i\theta)\right]^{1-Y_i},$$

where the inverse logit function is

$$\text{logit}^{-1}(a) = \exp(a)/[1 + \exp(a)]$$

corresponding to the logit function

$$\text{logit}(a) = \log[a/(1 + a)],$$

and θ is a column vector with the same number of components as X_i. This specification is known as the logistic regression specification, and as with linear regression, functions of the X_i can be used in place of the X_i with no essential change.

A new feature when using logistic regression as opposed to linear regression, however, is that the posterior distribution of θ no longer has a neat form for reasonable prior distributions. In fact, although the posterior distribution of θ for large enough n_1 is nearly normal, in many practical cases it is very far from normal, especially with modest n_1, many components in X, and data such that the fraction of ones in Y_{obs} is near zero or one.

Nevertheless, common practice uses the large-sample normal approximation assuming $\Pr(\theta) \propto \text{const}$ and thus approximates the posterior mean of

θ, $E(\theta|X, Y_{obs})$, by the maximum-likelihood estimate $\hat{\theta}$, defined by

$$\prod_{i \in obs} f(Y_i|X_i, \hat{\theta}) \geq \prod_{i \in obs} f(Y_i|X_i, \theta) \quad \text{for all } \theta$$

and the posterior variance of θ, $V(\theta|X, Y_{obs})$, by the negative inverse of the second derivative matrix of the log-posterior distribution at $\theta = \hat{\theta}$:

$$\hat{V}(\hat{\theta}) = -\left[\frac{\partial^2}{\partial\theta\partial\theta}\log \prod_{i \in obs} f(Y_i|X_i, \theta)\bigg|_{\theta=\hat{\theta}}\right]^{-1}.$$

Finding $\hat{\theta}$ requires iteration, although $V(\theta)$ is a straightforward function of θ. Using these approximations, the calculation of $\hat{\theta}$ and then $\hat{V}(\hat{\theta})$ defines the estimation task.

The imputation task is then

1. Draw θ from $N(\hat{\theta}, \hat{V}(\hat{\theta}))$, say θ_*.
2. For $i \in mis$, calculate $\text{logit}^{-1}(X_i\theta_*)$.
3. Draw n_0 independent uniform $(0,1)$ random numbers, u_i, $i \in mis$, and if $u_i > \text{logit}^{-1}(X_i\theta_*)$ impute $Y_i = 0$, otherwise impute $Y_i = 1$.

These steps are repeated for each of the m sets of imputations with new draws of random numbers. Section 5.5 illustrates the results of such calculations in a particular data set.

5.4. MONOTONE PATTERNS OF MISSINGNESS IN MULTIVARIATE Y_i

There are many practical cases in which creating multiple imputations with multivariate Y_i is only slightly more difficult than with univariate Y_i. Usually these cases involve special patterns of missingness in Y called monotone. In order to have a clean definition of monotone missingness, we must assume that all unobserved values for a unit included in the survey are to be regarded as missing values. Thus, we are essentially assuming that the sampling mechanism is such that if the ith unit is selected for inclusion in the survey, all components of Y_i are intended to be observed, that is, $I_i = (1, \ldots, 1)$ or $I_i = (0, \ldots, 0)$, and thus any unobserved values for a unit included in the sample are due to nonresponse. This assumption is no real restriction when creating imputations. Since we are assuming ignorable nonresponse, both unobserved values in Y_i due to the ignorable sampling

MONOTONE PATTERNS OF MISSINGNESS

Variables

	1	1	1	1	1	1	1
	1	1	1	1	1	1	0
Units	1	1	1	1	1	0	0
	1	1	1	0	0	0	0
	1	1	0	0	0	0	0

Figure 5.1. A monotone pattern of missingness, 1 = observed, 0 = missing.

mechanism, $Y_{i,exc}$, and unobserved values in Y_i due to ignorable nonresponse, $Y_{i,mis}$, are due to ignorable processes, and both can be imputed from the modeling of the distribution of (X, Y) needed to impute $Y_{i,mis}$; if desired, when analyzing the data, the imputations for $Y_{i,exc}$ can always be ignored.

Monotone Missingness in Y—Definition

Figure 5.1 displays a "monotone" or "nested" pattern of missingness in Y. The first component of Y is at least as observed as the second, which is at least as observed as the third component, and so on. Such a pattern of missingness, or a close approximation to it, is not uncommon in practice. For example in longitudinal surveys, subjects often drop out as the survey progresses, so that all units have all time 1 measurements, a subset of these have time 2 measurements, a subset of the units with time 2 measurements have time 3 measurements, and so on; Marini, Olsen, and Rubin (1980) provide a specific example with a missingness pattern that is close to monotone. Also, it is common that some units will answer all questions, another group of units will answer all questions except a collection of personal ones, such as those concerning income, and a final group will answer only background questions such as those about gender and family size. Monotone patterns are of particular interest here because they can be handled using only the tools developed for univariate Y_i such as presented in Examples 4.2–4.5 and 5.1–5.3.

The General Monotone Pattern—Description of General Techniques

In order to describe procedures for the general monotone pattern, it helps to have a notation to refer to a particular column of Y just as we use Y_i to refer to the ith row of Y. We let $Y_{[j]} = (Y_{1j}, \ldots, Y_{Nj})^T$. With a general monotone pattern, $Y_{[1]}$ is at least as observed as $Y_{[2]}$, which is at least as observed as $Y_{[3]}, \ldots,$ which is at least as observed as $Y_{[p]}$, as in Figure 5.1.

The general recommended procedure is as follows. The missing values of $Y_{[1]}$ are imputed from X ignoring the other components of Y; the missing values of $Y_{[2]}$ are imputed from $(Y_{[1]}, X)$ ignoring the other components of Y, and so on. Each of these imputation models can be independently applied using methods developed for univariate Y_i. Furthermore, the models used can vary in type. For example, the model used to impute $Y_{[1]}$ from X could be an implicit model such as an MV hot-deck (see Example 4.5), the model used to impute $Y_{[2]}$ from $X, Y_{[1]}$ could be an explicit linear regression model (see Example 5.1), and the model used to impute $Y_{[3]}$ from $(X, Y_{[1]}, Y_{[2]})$ could be an explicit logistic regression model (see Example 5.3).

We first illustrate the method with bivariate Y_i and implicit and explicit models, and then present two general results that justify the procedures we have described.

Example 5.4. Bivariate Y_i and an Implicit Imputation Model

Figure 5.2 provides an artificial example with nine units and a monotone missing data pattern; two imputed values were created for each missing value by a hot-deck procedure with ABB draws (see Example 4.4), and these are enclosed by parentheses. The procedure used was as follows. First, the missing value in $Y_{[1]}$ was imputed twice using fully observed X and ignoring the less observed $Y_{[2]}$; since $X_9 = 2$ and $X_5 = X_7 = X_8 = 2$, the fifth, seventh, and eighth units are possible donors for the ninth unit; two independent ABB draws from $Y_{51} = 1$, $Y_{71} = 2$, and $Y_{81} = 1$ happened to give 1 and 2 for the two imputed values of Y_{91}.

Next, the two missing values in $Y_{[2]}$, Y_{i2} for $i = 8$ and 9 were each imputed twice using a method for fully observed X and $Y_{[1]}$. In order to

i	X_i	Y_{i1}	Y_{i2}
1	1	1	1
2	1	1	0
3	1	0	0
4	1	1	1
5	2	1	2
6	1	1	2
7	2	2	1
8	2	1	(2, 2)
9	2	(1, 2)	(2, 1)

Figure 5.2. Artificial example illustrating hot-deck multiple imputation with a monotone pattern of missing data; parentheses enclose $m = 2$ imputations.

create the first set of imputed Y_{i2}, treat the first set of imputed Y_{i1} as real; that is, suppose $Y_{91} = 1$. Then there are two units that match unit 8 on $(X, Y_{[1]})$—the fifth and ninth, but only the fifth has $Y_{[2]}$ observed, so that its value, $Y_{52} = 2$, is imputed for Y_{82}; similarly, there are two units that match the ninth unit's values X and $Y_{[1]}$—the fifth and the eighth, but only the fifth has $Y_{[2]}$ observed so that its value $Y_{52} = 2$ is imputed for Y_{92}. In order to create the second set of imputed Y_{i2}, treat the second set of imputed Y_{i1} as real, that is, suppose $Y_{91} = 2$. Then the eighth unit has only unit 5 as a matching donor for $Y_{[2]}$ and so once again the imputed value of Y_{82} is 2. The ninth unit now has $X_9 = 2$, $Y_{91} = 2$, and so matches unit 7, which donates 1 as the imputed value of Y_{92}.

The imputations could have been performed in a different way. First, impute *one* value for each missing Y_{i1}, and treating them as real, impute one value for each missing Y_{i2}; this creates the first set of imputations. Second, repeat both steps with new random draws.

Example 5.5. Bivariate Y_i with an Explicit Normal Linear Regression Model

A more formal example occurs with $Y_i = (Y_{i1}, Y_{i2})$ having a bivariate normal linear regression on X_i, with Y_{i1} observed on n_1 units and Y_{i2} observed on $n_2 \le n_1$ units in a monotone pattern. Corresponding to the monotone pattern, we specify the joint distribution of (Y_{i1}, Y_{i2}) given X_i by first specifying the conditional distribution of Y_{i1} given (X_i, θ) as

$$N(X_i\beta_1, \sigma_1^2),$$

and then the conditional distribution of Y_{i2} given (Y_{i1}, X_i, θ) as

$$N(\gamma Y_{i1} + X_i\beta_2, \sigma_2^2),$$

where β_1 and β_2 are column vectors of q components, and γ, σ_1^2 and σ_2^2 are scalars, with $\theta = (\beta_1, \beta_2, \gamma, \log \sigma_1, \log \sigma_2)$ and $\Pr(\theta) \propto \text{const}$.

First, ignore Y_{i2} and impute the missing values of Y_{i1} using the fully observed X and applying results for the normal linear model given in Example 5.1, where notationally Y_i is replaced by Y_{i1}, β is replaced by β_1, σ^2 by $\sigma_1^2, \ldots$, and *obs* by *obs*[1]—the set of units with $Y_{[1]}$ observed. Thus, parameters of the regression of $Y_{[1]}$ on X, $\theta_1 = (\beta_1, \log \sigma_1)$, are estimated from the units with $Y_{[1]}$ observed, and missing $Y_{[1]}$ values are imputed by first drawing from the posterior distribution of these parameters, and then using X_i to predict missing Y_{i1}. Suppose two sets of missing Y_{i1} have been imputed.

Now treat the first set of imputed Y_{i1} as real and impute one set of missing Y_{i2} using the results in Example 5.1 where the regression is that of Y_{i2} on (Y_{i1}, X_i), and its parameters are estimated using the units with Y_{i2} observed. Thus, the notation of Example 5.1 is changed so that Y_i is replaced by Y_{i2}, β is replaced by (γ, β_2), σ^2 by $\sigma_2^2,\ldots$, and obs by $obs[2]$ — the set of units with Y_{i2} observed. Next, treat the second set of imputed Y_{i1} as real and impute a second set of missing Y_{i2} using the same procedure but with new random draws.

The posterior distribution of $\theta_2 = (\gamma, \beta_2, \log \sigma_2)$ does not involve any imputed Y_{i1} since the monotone pattern implies that all units with Y_{i2} observed have Y_{i1} observed. Thus, the posterior distribution of θ_2 treating the first set of imputed Y_{i1} as real is identical to the posterior distribution of θ_2 treating the second set of imputed Y_{i1} as real. Imputed values of Y_{i2} can, however, depend on imputed values of Y_{i1} since units with both Y_{i2} and Y_{i1} missing will use the imputed Y_{i1} values in calculating the conditional mean of the Y_{i2} value to be imputed.

If $\theta_1 = (\beta_1, \log \sigma_1)$ and $\theta_2 = (\gamma, \beta_2, \log \sigma_2)$ had been dependent *a priori*, they would be dependent *a posteriori*, and then the simple procedure just described would not have been appropriate. Also, without a monotone pattern of missingness, the procedure would not have been clearly defined. In general, for the type of procedure we have described to be entirely applicable, we need a *monotone-distinct* structure.

Monotone-Distinct Structure

Two conditions must hold for the missingness-modeling structure to be monotone-distinct. First, the missing data pattern must be monotone with $Y_{[1]}$ at least as observed as $Y_{[2]}$, which is at least as observed as $Y_{[3]},\ldots$, which is at least as observed as $Y_{[p]}$. Second, defining p functions of θ by the factorization

$$f(Y_i|X_i, \theta) = f_1(Y_{i1}|X_i, \theta_1) f_2(Y_{i2}|X_i, Y_{i1}, \theta_2)$$
$$\cdots f_p(Y_{ip}|X_i, Y_{i1},\ldots, Y_{i,p-1}, \theta_p), \quad (5.4.1)$$

the $\theta_1,\ldots, \theta_p$ must be distinct in the sense that they are *a priori* independent

$$\Pr(\theta) = \prod_{j=1}^{p} \Pr(\theta_j). \quad (5.4.2)$$

The term *distinct* was defined in Rubin (1974) to include both (5.4.2) when prior distributions are defined and a factorized parameter space when they are not.

MONOTONE PATTERNS OF MISSINGNESS

Some extra notation is useful for stating and proving results precisely. First, let $obs[j] = \{i | I_{ij} R_{ij} = 1\}$, that is, $obs[j]$ is the set of units with $Y_{[j]}$ observed. Thus, the missingness is monotone if

$$obs[1] \supseteq obs[2] \supseteq \cdots \supseteq obs[p] \quad (5.4.3)$$

so that if $i \in obs[j]$ then $i \in obs[j-1]$ for $j = 2, \ldots, p$. Second, let f_{ij} be the jth factor on the right-hand side of (5.4.1):

$$f_{ij} = f_j(Y_{ij} | X_i, Y_{i1}, \ldots, Y_{i,j-1}, \theta_j),$$

so that

$$f(Y_i | X_i, \theta) = \prod_{j=1}^{p} f_{ij}. \quad (5.4.4)$$

Generally f_{ij} is a function of Y_{ij}, X_i, θ_j, and all Y_{it} such that $t < j$, and by (5.4.3), if $i \in obs[j]$, then f_{ij} is functionally free of any unobserved Y_{ij}.

Result 5.4. The Estimation Task with a Monotone-Distinct Structure

Suppose the missingness-modeling structure is monotone-distinct. Then the estimation task is equivalent to a series of p independent estimation tasks, each with univariate Y_i: the jth task estimates the conditional distribution of $Y_{[j]}$ given the more observed variables $(X, Y_{[1]}, \ldots, Y_{[j-1]})$ using the set of units with $Y_{[j]}$ observed, $obs[j]$. Explicitly, the claim is that $\theta_1, \ldots, \theta_p$ are *a posteriori* independent with

$$\Pr(\theta_j | X, Y_{obs}) = \frac{\prod_{i \in obs[j]} f_{ij} \Pr(\theta_j)}{\int \prod_{i \in obs[j]} f_{ij} \Pr(\theta_j) \, d\theta_j}, \quad (5.4.5)$$

where (5.4.5) is formally equivalent to (5.2.12), the posterior distribution of $\theta_{Y|X}$ with univariate Y_i.

The proof of Result 5.4 is straightforward. First, by Bayes's theorem, and the definition of θ and $\Pr(\theta)$ in (5.2.14),

$$\Pr(\theta | X, Y_{obs}) = \frac{\Pr(Y_{obs} | X, \theta) \Pr(\theta)}{\int \Pr(Y_{obs} | X, \theta) \Pr(\theta) \, d\theta}. \quad (5.4.6)$$

Now

$$\Pr(Y_{obs} | X, \theta) = \int \Pr(Y | X, \theta) \, dY_{nob} \quad (5.4.7)$$

where by (5.2.15) and (5.4.4)

$$\Pr(Y|X, \theta) = \prod_{j=1}^{p} \prod_{i=1}^{N} f_{ij}. \qquad (5.4.8)$$

Also

$$dY_{nob} = \prod_{j=1}^{p} \prod_{i \in nob[j]} dY_{ij} \qquad (5.4.9)$$

where $nob[j] = \{i | I_{ij}R_{ij} = 0\}$, $nob[j] \cup obs[j] = \{1, \ldots, N\}$ for all j. Substituting (5.4.9) and (5.4.8) into (5.4.7) gives

$$\Pr(Y_{obs}|X, \theta) = \int \cdots \int \left[\prod_{j=1}^{p} \prod_{i=1}^{N} f_{ij} \right] \left[\prod_{j=1}^{p} \prod_{i \in nob[j]} dY_{ij} \right].$$

Rearranging this expression gives

$$\Pr(Y_{obs}|X, \theta) = \int \cdots \int \left[\prod_{j=1}^{p} \prod_{i \in obs[j]} f_{ij} \right] \left[\prod_{j=1}^{p} \prod_{i \in nob[j]} f_{ij} \, dY_{ij} \right]. \qquad (5.4.10)$$

But by (5.4.3) and the observation following (5.4.4), the first bracketed factor in (5.4.10) is functionally free of any unobserved Y_{ij}. Hence the factor can be brought outside the multiple integral in (5.4.10), and the remaining integral is one. Thus

$$\Pr(Y_{obs}|X, \theta) = \prod_{j=1}^{p} \prod_{i \in obs[j]} f_{ij}. \qquad (5.4.11)$$

Up until now we have not appealed to the distinctness of the θ_j. Using (5.4.2) and (5.4.11) in (5.4.6) yields

$$\Pr(\theta|X, Y_{obs}) = \prod_{j=1}^{p} \frac{\prod_{i \in obs[j]} f_{ij} \Pr(\theta_j)}{\int \prod_{i \in obs[j]} f_{ij} \Pr(\theta_j) \, d\theta_j},$$

which proves the primary claim of Result 5.4.

The formal equivalence of (5.4.5) and (5.2.12) follows from noting that $f_j(\cdot|\cdot)$ plays the role of $f_{Y|X}(\cdot|\cdot)$ in (5.2.12), $Y_{[j]}$ plays the role of Y, $(X, Y_{[1]}, \ldots, Y_{[j-1]})$ plays the role of X when $j > 1$ (when $j = 1$, X plays the role of X), θ_j plays the role of $\theta_{Y|X}$, and $obs[j]$ plays the role of obs.

Result 5.5. The Imputation Task with a Monotone-Distinct Structure

Suppose that the missingness-modeling structure is monotone-distinct. Then the imputation task is equivalent to a sequence of p independent imputation tasks, each with univariate Y_j: the jth task independently imputes the missing values of $Y_{[j]}$ using their conditional distributions given θ and the observed values $X_i, Y_{i1}, \ldots, Y_{i,p-1}$. Explicitly, the claim is that the posterior distribution of Y_{mis} given θ is

$$\Pr(Y_{mis}|X, Y_{obs}, \theta) = \prod_{j=1}^{p} \prod_{i \in mis[j]} f_{ij} \qquad (5.4.12)$$

where $mis[j] = \{i | I_{ij} = 1 \text{ and } R_{ij} = 0\}$ = the units missing variable $Y_{[j]}$, and (5.4.12) is the product of p conditional distributions, each of which is formally equivalent to (5.2.13), the posterior distribution of Y_{mis} given $\theta_{Y|X}$ with univariate Y_i.

The proof of Result 5.5 is quite simple. By Bayes's theorem,

$$\Pr(Y_{mis}|X, Y_{obs}, \theta) = \frac{\Pr(Y_{mis}, Y_{obs}|X, \theta)}{\Pr(Y_{obs}|X, \theta)}. \qquad (5.4.13)$$

Since we are assuming $I_i = (1, \ldots, 1)$ or $(0, \ldots, 0)$ and $Y_{inc} = (Y_{mis}, Y_{obs})$, from (5.4.8)

$$\Pr(Y_{mis}, Y_{obs}|X, \theta) = \prod_{j=1}^{p} \prod_{i \in inc} f_{ij}. \qquad (5.4.14)$$

Substituting (5.4.11) and (5.4.14) into (5.4.13) gives

$$\Pr(Y_{mis}|X, Y_{obs}, \theta) = \prod_{j=1}^{p} \frac{\prod_{i \in inc} f_{ij}}{\prod_{i \in obs[j]} f_{ij}},$$

which equals (5.4.12) since $inc = obs[j] \cup mis[j]$. A careful look at (5.4.11), (5.4.13), and (5.4.14) reveals that the validity of (5.4.12) does not require the independence of the θ_j in (5.4.3).

To establish the formal correspondence between each factor in (5.4.12) and (5.2.13), note that for the jth factor, $f_j(\cdot|\cdot)$ plays the role of $f_{Y|X}(\cdot|\cdot)$ in (5.2.13), $Y_{[j]}$ plays the role of Y, $(X, Y_{[1]}, \ldots, Y_{[j-1]})$ plays the role of X for $j > 1$ (for $j = 1$, X plays the role of X), θ_j plays the role of $\theta_{Y|X}$, and $mis[j]$ plays the role of mis. The first factor is the conditional distribution of the missing values in $Y_{[1]}, Y_{i1}, i \in mis[1]$, given X and θ_1; the jth factor for $j > 1$ is the conditional distribution of the missing values in $Y_{[j]}$, $\{Y_{ij}|\, i \in mis[\,j\,]\}$, given θ_j and the missing and observed values in $X, Y_{[1]}, \ldots, Y_{[j-1]}$.

5.5. MISSING SOCIAL SECURITY BENEFITS IN THE CURRENT POPULATION SURVEY

The example we use to illustrate monotone missingness concerns missing social security benefits for a number of individuals surveyed during the Census Bureau's March 1973 Current Population Survey (CPS). This survey is of particular interest because values for social security benefits are available from administrative records for both respondents and nonrespondents. Although the administrative values are not necessarily the same as survey responses, they do provide an objective procedure for studying the results of imputation procedures. The outcome is bivariate: $Y_{i1} = 1$ indicates some benefits were received, whereas $Y_{i1} = 0$ indicates that no benefits were received; Y_{i2} provides the amount of benefits in dollars, where $Y_{i2} = 0$ if $Y_{i1} = 0$. Nonresponse is such that some units divulge neither recipiency status (Y_{i1}) nor amount (Y_{i2}), others respond about recipiency status but not amount, thereby providing Y_{i1} but not Y_{i2}, while others fully respond by giving an exact amount for benefits and thereby provide both Y_{i1} and Y_{i2}. Thus, the nonresponse pattern on the outcome Y is monotone. The model we consider assumes distinctness in the underlying associated parametric structure, and thus we apply the results of Section 5.4, which hold with a monotone-distinct structure.

The CPS–IRS–SSA Exact Match File

The CPS is a multistage, stratified clustered sample conducted monthly by the Census Bureau with over 50,000 households. Although a multipurpose survey, the principal goal is to estimate the labor force status of noninstitutionalized civilians at least 14 years of age. The March CPS interview always includes a series of income questions designed to ascertain all sources of income during the preceding calendar year. Nonresponse rates on these income questions are rather high (i.e., 5–20% depending on the particular question). The CPS–SSA–IRS Exact Match File (Aziz, Kilss, and Scheuren, 1978) is the March 1973 CPS supplemented with administrative records from both the Social Security Administration (SSA) and the Internal Revenue Service (IRS). It was created by combining CPS survey data with SSA benefit and earnings records and IRS federal tax return records, using social security numbers of individuals in the March 1973 CPS to match records in the SSA and IRS files. The resultant exact match file combines the probability sampling and nonresponse structure of the CPS with the essentially complete data on the income and benefits from the administrative records of the IRS and SSA.

The Reduced Data Base

The exploratory study of missing social security benefits reported here was jointly conducted by T. Herzog, C. Lancaster, F. Scheuren, and D. Rubin, and relevant material is reported in more detail in Herzog (1980), Herzog and Lancaster (1980), Herzog and Rubin (1983), Lancaster (1979), and Oh and Scheuren (1980). In order to have a clean data base from which to study missing social security benefits, attention was restricted to the 1058 males in the March 1973 CPS who were at least 62 years of age as of December 1972 and satisfied other selection criteria documented in Herzog and Rubin (1983). Of these 1058 individuals, 999 reported CPS OASDI (old age–survivor and disability income) benefit amounts, and all remaining 59 reported neither amount nor recipiency status in the CPS.

The data analyses used response–nonresponse on the CPS OASDI questions and actual values for OASDI from administrative records to create a data set with the nonresponse pattern of the CPS, but income data from administrative records for all individuals. Hence, the resulting exploratory data base had for each of the 1058 individuals: OASDI benefits from administrative records (Y_{i1}, Y_{i2}), response–nonresponse on the CPS OASDI questions (R_{i1}, R_{i2}), as well as demographic variables from the CPS (e.g., age, years of school, ethnicity, urban/rural, marital status, weeks worked during calendar year), which were considered to be the fully observed X variables. Treating the administrative data for nonrespondents as missing enabled the accuracy of the imputations for nonrespondents based on respondent data to be compared to actual data from nonrespondents.

Since the nonresponse rate for SSA benefits in this data base is relatively low ($59/1058 = 5.6\%$), we would not expect inferences for mean OASDI benefits in the population to be greatly affected by the particular imputation procedure. Nevertheless, this situation can be used to illustrate imputation procedures appropriate with monotone-distinct missingness as well as to evaluate the procedures with respect to accuracy of prediction and interval estimation.

The Modeling Task

Because of the monotone missingness for (Y_{i1}, Y_{i2}), two independent models were specified to create a monotone-distinct missingness structure. The first model is for Y_{i1} given predictor variables X_i, and the second model is for Y_{i2} given X_i and Y_{i1}.

The model for recipiency status, Y_{i1}, given X_i is a logistic regression model, introduced here in Example 5.3:

$$\text{Prob}\{Y_{i1} = 1 | X_i, \lambda\} = \exp(X_i\lambda)/[1 + \exp(X_i\lambda)] = \text{logit}^{-1}(X_i\lambda)$$

where λ is the regression coefficient parameter specified to have prior distribution proportional to a constant. The specific model, described in Lancaster (1979), included interactions among the X variables as well as indicator variables (e.g., ≥ 72 years old).

The model for OASDI amount (Y_{i2}) given recipiency status (Y_{i1}) and X_i is as follows: if $Y_{i1} = 0$, then $Y_{i2} = 0$ with probability 1 for all X_i, whereas if $Y_{i1} = 1$, then $\log(Y_{i2})$ has a normal linear regression on X_i:

$$\left(\log(Y_{i2}) | Y_{i1} = 1, X_i, \beta, \sigma^2\right) \sim N(X_i\beta, \sigma^2),$$

where the prior distribution on $(\beta, \log \sigma)$ is proportional to a constant. As with the logistic regression model, X_i included constructed predictor variables (e.g., age squared) in addition to the raw variables.

The Estimation Task

By Result 5.4, there are two independent estimation tasks for this example, one for each of the independent models.

The first estimation task is to obtain the posterior distribution relevant to the logistic regression of Y_{i1} on X_i using the units with complete data on (Y_{i1}, X_i). There were 999 such units, and the resultant posterior distribution of λ was approximated by a multivariate normal distribution with mean given by the maximum-likelihood estimate (posterior mode) and variance–covariance matrix given by minus the inverse of the second derivative of the log likelihood (log posterior) at this maximum. This procedure is justifiable when sample sizes are large; it is not very satisfactory when sample sizes are small because the likelihood function for logistic regression parameters can be distinctly nonnormal with small samples, as discussed in Example 5.3.

The second estimation task is to obtain the posterior distribution relevant to the regression of Y_{i2} on (Y_{i1}, X_i) using the units with data on (Y_{i2}, Y_{i1}, X_i). There were 999 such units, and the resultant posterior distribution of (β, σ^2) was the standard one defined by least-squares statistics described in Example 5.1.

The Imputation Task

One hundred imputed values of (Y_{i1}, Y_{i2}) were created under this logistic regression–linear regression model. The imputation task was thus repeated 100 times, each time using a new independent draw of parameters from their posterior distribution. Each imputation task was implemented following Result 5.5, using a sequence of two independent imputation tasks each designed for univariate Y_i.

The first univariate imputation task is for Y_{i1}. First, draw λ from its approximating normal posterior distribution defined in the estimation task, say λ_*. For units missing Y_{i1}, draw independent uniform random numbers on $(0,1)$, u_i, $i \in mis$, and calculate $X_i\lambda_*$; if $u_i > \exp(X_i\lambda_*)/[1 + \exp(X_i\lambda_*)]$, impute $Y_{i1} = 0$, otherwise impute $Y_{i1} = 1$.

The second univariate imputation task is for Y_{i2}. First, draw (β, σ^2) from its posterior distribution defined in the estimation task, say (β_*, σ_*^2). For each unit missing Y_{i2}, examine Y_{i1}: If $Y_{i1} = 0$, impute $Y_{i2} = 0$; if $Y_{i1} = 1$, draw an independent $N(0,1)$ deviate, say z_i, and impute $Y_{i2} = \exp(X_i\beta_* + z_i\sigma_*)$.

Tables 5.1 and 5.2 summarize the results of the imputation task for all 59 respondents, first for those 62–71 and then for those over 72 years of age, since the administrative rules allowed benefits for those over 72 even if no Social Security benefits had been paid. In addition to the "explicit model values" obtained under the previously described explicit modeling procedure, Tables 5.1 and 5.2 also give two values obtained from a multiple-imputation version of the Census Bureau's hot-deck approach, modified to impute administrative values rather than CPS survey responses (Herzog and Rubin, 1983, provide details). The mean administrative value for the 999 respondents is 1455.

Results Concerning Absolute Accuracies of Prediction

Table 5.3 summarizes the accuracies of prediction for hot-deck and model-based imputation methods across the 59 nonrespondents, accuracy in the sense of how well the imputation methods reproduce the actual administrative values of OASDI benefits. Using either mean absolute deviation or root-mean-squared deviation, it is easy to see that estimates based on means of multiply-imputed values tend to be more accurate than estimates based on one imputed value, and also that the model-based estimates tend to be better than the hot-deck estimates. The conclusion is that, in this example, the explicit model-based posterior mean provided by 100 multiple imputations repeatedly drawn from the posterior predictive distribution of the missing values is the best estimate with respect to either criterion.

TABLE 5.1. Multiple imputations of OASDI benefits for nonrespondents 62–71 years of age.[a]

Nonrespondent Number	Admin. Value	Hot-Deck Values		Explicit Model Values			Mean of All 100	Standard Deviation of 100
		First Value	Second Value	First Value	Second Value	Percent Recipiency		
1	2027	1845	2613	0	2438	72	1396	967
2	1047	1352	1548	0	0	51	835	867
3	2095	2613	2239	1692	1698	67	769	604
4	2112	2703	2685	1595	1753	87	1688	816
5	1947	2314	0	2142	1945	60	1066	930
6	2242	1870	815	2221	1942	89	1952	860
7	2341	2161	2025	0	2711	79	1711	1025
8	2631	1877	1451	2852	2252	84	1940	1007
9	2597	0	2314	2531	1511	85	2144	1066
10	0	0	0	0	2697	73	1737	1210
11	0	0	0	884	917	66	702	547
12	561	735	887	1356	1093	78	1463	899
13	2662	1778	666	1096	735	73	676	470
14	992	1546	993	1450	2090	87	1848	913
15	2204	1274	1035	1612	1537	83	1562	844
16	0	0	0	821	0	63	416	347
17	1928	1378	1010	1452	2691	90	1934	811
18	2392	1718	2660	1319	0	61	1102	957
19	2217	2018	1994	2492	1400	81	1869	1057
20	2685	2085	1920	2938	2680	79	2021	1185
21	1491	1279	1102	0	1009	55	897	875
22	0	2604	0	1535	0	64	712	578
23	1056	2058	1757	1411	1743	51	787	815
24	2343	2027	965	1472	1468	80	1477	861
25	0	2651	609	1782	1873	87	1437	686
26	1930	3103	2161	1846	1477	67	893	708
27	0	0	0	0	1007	73	824	559
28	2169	2210	2314	2464	2355	79	1830	1065
29	2597	965	2027	1851	2271	78	1691	1014
30	848	1718	2315	1694	0	85	1917	945
31	0	2411	2703	2477	2592	89	2157	941
32	2450	1438	1313	2074	1972	84	1946	1034
33	0	1568	1597	1820	2263	81	2125	1203
34	887	1031	1687	2725	1572	81	1713	951
35	410	1641	1990	0	1027	62	765	658
36	0	2631	2597	2110	1589	57	985	924
37	1016	0	0	0	920	70	548	398
38	1445	2199	2203	2111	1521	57	1119	1031
39	1774	2090	2239	1628	2537	74	1634	1091
40	2505	1313	1438	0	2683	84	1829	936
41	1846	2523	2218	0	2883	84	1969	1047
Mean:	1450	1627	1466	1401	1631	74	1417	

[a] *Source*: Table 9 in Herzog and Rubin (1983).

TABLE 5.2. Multiple imputations of OASDI benefits for nonrespondents over 72 years of age.[a]

Nonrespondent Number	Admin. Value	Hot-Deck Values		Explicit Model Values				
		First Value	Second Value	First Value	Second Value	Percent Recipiency	Mean of All 100	Standard Deviation of 100
1	1489	2085	2257	1401	1181	99	1555	421
2	2103	609	645	1650	1417	95	1720	596
3	2027	1968	2142	975	1020	100	1348	338
4	993	2059	2375	1858	2223	96	1596	517
5	0	2161	2025	3360	2656	81	2497	1445
6	2199	2296	2432	0	1270	68	852	637
7	2487	1968	1839	2276	0	77	1797	1120
8	2375	2271	2363	1886	0	98	1949	561
9	1419	609	666	2782	3343	72	1773	1258
10	1854	0	2085	1693	2001	100	1939	469
11	2161	1603	1864	1498	2772	98	1873	567
12	2957	605	799	2256	0	73	1725	1170
13	1187	2774	2348	1102	1779	100	1591	373
14	1835	791	797	1860	1586	82	1812	1001
15	887	1344	1609	2091	1317	95	1546	544
16	2375	1864	1603	2804	2392	99	1950	528
17	2384	0	2354	1689	1699	84	1756	920
18	2239	887	735	1165	1074	95	1261	429
Mean:	1832	1439	1719	1797	1541	90	1697	

[a]*Source*: Table 9 in Herzog and Rubin (1983).

TABLE 5.3. Accuracies of imputation methods with respect to mean absolute deviation (MAD) and root mean squared deviation (RMS).[a]

Nonrespondent Group			Hot-Deck		Model-Based		
			First Value	Mean of First Two Values	First Value	Mean of First Two Values	Mean of all 100 Values
62–71 years		MAD	826	786	908	753	776
		RMS	1133	1001	1175	967	934
72 + years		MAD	1056	937	840	940	654
		RMS	1296	1122	1168	1212	868

[a]*Source*: Table 10 in Herzog and Rubin (1983).

Inferences for the Average OASDI Benefits for the Nonrespondents in the Sample

In order to judge whether the discrepancies between imputed and actual OASDI benefits indicate failures of the imputation models, we compare the actual average OASDI benefits for the 59 nonrespondents in the sample based on administrative records with the inference for this average implied by each imputation method. If the values for the nonrespondents were known, their average would be known. Thus, the complete-data inference for this sample average has zero variance, and as a result, the single-imputation inferences estimate this average by the average of the imputed values with zero standard error, as given in Table 5.4. Inferences with $m \geq 2$ imputations are obtained by treating the multiple imputations under each method as repeated imputations under one model and applying the repeated-imputation procedure for interval estimation outlined in Section 3.1 and derived in Section 3.3. Table 5.4 summarizes results. Each repeated-imputation estimate is the average of the m complete-data averages obtained under that method. The standard error associated with each repeated-imputation estimate is the square root of $(1 + m^{-1})$ times the sample variance of the m complete-data averages across the multiple imputations, and the degrees of freedom for the associated t reference distribution is $(m - 1)$—recall that the within-variance component is zero because the estimand is the average OASDI benefits for the nonrespondents in the *sample*.

Clearly, the single-imputation methods imply unrealistic posterior distributions for the nonrespondents. All three multiple-imputation procedures considered imply posterior distributions that comfortably cover the true average OASDI benefits for the nonrespondents in the sample. Consequently, there is no real evidence in Table 5.4 to suggest that any of the

TABLE 5.4. Comparison of estimates (standard errors) for mean OASDI benefits implied by imputation methods for nonrespondent groups in the sample.

	Hot-Deck		Model-Based		
Nonrespondent Group: True mean	First Value	Two Values	First Value	Two Values	100 Values
62–71 years: 1450	1627 (0)	1547 (139)	1401 (0)	1516 (198)	1417 (201)
72 + years: 1832	1439 (0)	1579 (242)	1797 (0)	1669 (222)	1697 (196)

three multiple-imputation methods is deficient, and thus, in particular, no evidence to suggest a nonignorable response mechanism since only ignorable models have been considered. Further checks should be made, however, before definitely accepting this conclusion. For example, the analysis presented in Table 5.4 could be done for other subgroups defined by X variables besides 62–71 and 72+ years old.

Results on Inferences for Population Quantities

Now consider inferences for population quantities, such as the mean OASDI benefits for all nonrespondents in the population 62–71 and 72+ years old, and all individuals 62+ years old. For simplicity, assume that the standard interval would be appropriate if there were observed values from nonrespondents, and again apply the repeated-imputation procedure outlined in Section 3.1, where the estimate is the average of the m complete-data estimates, and the squared standard error associated with the repeated-imputation estimate is the sum of two components: the average within imputation variance and $(1 + m^{-1})$ times the between-imputation variance (which is zero for single-imputation methods). The degrees of freedom for the t reference distribution when $m > 1$ are $(m - 1)(1 + r_m^{-1})^2$ where r_m is the relative increase in conditional variance due to nonresponse, given by $(1 + m^{-1})$ times the ratio of between to within variance.

Table 5.5 presents results derived from Herzog and Rubin (1983). Relative to the intervals based on 100 model-based imputations, the other methods tend to give standard errors that are too small. The effect is

TABLE 5.5. Comparison of estimates (standard errors) for mean OASDI benefits implied by imputation methods for groups in the population.

	Hot-Deck		Model-Based		
Population Group[a]	First Value	Two Values	First Value	Two Values	100 Values
All nonrespondents 62–71	1627 (134)	1547 (194)	1401 (146)	1516 (242)	1417 (257)
All nonrespondents 72+	1439 (200)	1579 (302)	1797 (181)	1669 (301)	1697 (281)
All individuals 62+	1548 (27.1)	1547 (27.1)	1545 (27.1)	1547 (27.4)	1544 (27.5)

[a]As restricted by selection criteria.

relatively serious when using single-imputation methods to draw inferences about the mean OASDI benefits for nonrespondents in the population, in one case underestimating standard errors by nearly 50%.

This result is not surprising since so much information is missing for the nonrespondent groups. Using equation (3.1.10) and the standard errors based on 100 model-based imputations given in Table 5.4 and 5.5 to estimate the between and total components gives 62% missing information for the estimand "population mean for nonrespondents 62–71," and 49% missing information for the estimand "population mean for nonrespondents 72+ ."

Although estimands for nonrespondent groups are diagnostically interesting, usually the focus is on estimands defined solely by values of outcome variables Y and covariates X. In fact, the theoretical development in earlier chapters was for an estimand $Q = Q(X, Y)$, which did not involve R. The estimand "mean OASDI benefits for all individuals 62+ " is not greatly affected by nonresponse because nonrespondents comprise less than 6% of the population of individuals 62+ years old. Using the standard errors in Table 5.5 for "first value" and "100 values" to estimate the within and total components gives 3% as the fraction of information missing due to nonresponse for the mean of all individuals 62+ . This value is consistent with the intuitive "missing information for all individuals = fraction of individuals who are nonrespondents (6%) × fraction of information missing for nonrespondents (50%)."

5.6. BEYOND MONOTONE MISSINGNESS

Unfortunately, in many practical situations the pattern of missing data created by nonresponse is not monotone, and thus the results in Section 5.4 for a monotone-distinct missingness-modeling structure cannot be directly applied. Nevertheless, special cases exist where simple extensions of the techniques for univariate Y_i, which have already been discussed, can be directly applied. In particular, these cases are those in which the likelihood factors into pieces with distinct parameters such as those described by Rubin (1974). More generally, however, methods explicitly designed to handle multivariate Y_i must be employed.

Two Outcomes Never Jointly Observed—Statistical Matching of Files

File matching missingness is one of those special nonmonotone patterns for which univariate techniques can be applied. Suppose n values of $Y_i = (Y_{i1}, Y_{i2})$ were included in the sample but due to ignorable nonresponse, Y_{i1}

BEYOND MONOTONE MISSINGNESS

is observed for only n_1 units, Y_{i2} is observed for only n_2 units, where $n_1 + n_2 = n$, and Y_{i1} and Y_{i2} are never jointly observed. The likelihood in this case factors into two pieces: Y_{i1} given X_i for units with Y_{i1} observed and Y_{i2} given X_i for units with Y_{i2} observed. At first glance, this situation may look rather strange: Because Y_{i1} and Y_{i2} are never jointly observed, there are no data available to directly estimate the parameters of conditional association between Y_{i1} and Y_{i2} given X_i. Nevertheless, this structure is the one that exists in the statistical matching of files (Okner, 1974; Rodgers, 1984; Rubin, 1983e, 1986; Sims, 1972; Woodbury, 1983). One file has Y_{i1} and X_i recorded, another file has Y_{i2} and X_i recorded, and the desire is to have one file with Y_{i1}, Y_{i2}, and X_i observed for all units. In order to do any imputation, some assumption must be made about the conditional association between Y_{i1} and Y_{i2} given X_i. Standard methods implicitly or explicitly assume conditional independence.

Within the multiple-imputation framework it is not necessary to assume conditional independence or any other specific choice for the parameters of conditional association, because each set of imputations can be made for a difference choice of parameters of conditional association. If the values of these parameters across the imputations are considered to have been drawn from one prior distribution of the parameters, then the multiple imputations are repeated imputations under one model for nonresponse. If the values for these parameters are simply different possibilities, then the multiple imputations reveal sensitivity to different models for nonresponse. A simple normal example can be used to illustrate the essential ideas.

Example 5.6. Two Normal Outcomes Never Jointly Observed

Assume the file matching structure just described where for simplicity there are no X variables and (Y_{i1}, Y_{i2}) are modeled as i.i.d. normal:

$$\left(Y_{i1}, Y_{i2} | \mu_1, \mu_2, \sigma_1^2, \sigma_2^2, \rho\right) \sim N\left((\mu_1, \mu_2), \begin{pmatrix} \sigma_1^2 & \rho\sigma_1\sigma_2 \\ \rho\sigma_1\sigma_2 & \sigma_2^2 \end{pmatrix}\right),$$

where

$$\Pr(\mu_1, \mu_2, \log \sigma_1, \log \sigma_2 | \rho) \propto \text{const},$$

and $\Pr(\rho)$ is to be specified. Results in Rubin (1974) show that from the n_j observations of Y_{ij}, $j = 1, 2$, we obtain the usual posterior distribution for μ_j and σ_j^2: μ_j given σ_j^2 is $N(\hat{\mu}_j, \sigma_j^2/n_j)$ where $\hat{\mu}_j = \Sigma_{obs[j]} Y_{ij}/n_j$, and σ_j^2 is $\Sigma_{obs[j]}(Y_{ij} - \hat{\mu}_j)^2$ times an inverted χ^2 random variable on $n_j - 1$ degrees of freedom. The posterior distribution of ρ equals its prior distribution.

Suppose we have drawn values of $\theta = (\mu_1, \mu_2, \log \sigma_1, \log \sigma_2, \rho)$ from its posterior distribution; the procedure is obvious and familiar except that ρ is drawn from its prior, which may put all probability at one value (e.g., 0 or 0.5 or 1), or may specify, for example, that $\log[(1 + \rho)/(1 - \rho)] \sim N(0, K)$ for some fixed K. In any case, having drawn values of θ, say θ_*, the $n - n_1 = n_2$ missing Y_{i1} are drawn from their regression on observed Y_{i2}:

$$Y_{i1} = \mu_{1*} + \rho_* \sigma_{1*}(Y_{i2} - \mu_{2*})/\sigma_{2*} + z_i \sigma_{1*}\sqrt{1 - \rho_*^2},$$

where the z_i are i.i.d. standard normal deviates. Similarly, the $n - n_2 = n_1$ missing Y_{i2} are drawn from their regression on observed Y_{i1}:

$$Y_{i2} = \mu_{2*} + \rho_* \sigma_{2*}(Y_{i1} - \mu_{1*})/\sigma_{1*} + z_i \sigma_{2*}\sqrt{1 - \rho_*^2},$$

where z_i are i.i.d. standard normal deviates. A new draw of θ generates a new imputed value of Y_{mis}.

In many such cases, it may be of interest to investigate sensitivity to ρ rather than assume one prior distribution for it. If so, then several draws of $(\mu_1, \mu_2, \log \sigma_1, \log \sigma_2)$ at each of several values of ρ should be made, for example, two draws with $\rho = 0$, two draws with $\rho = 0.1$ and two draws with $\rho = 0.8$.

The extension of the above technique to cases with X variables is conceptually immediate: μ_j is replaced by $X_i \beta_j$, $(\log \sigma_1, \log \sigma_2, \rho)$ are interpreted as partial (or conditional) parameters given X_i, and the missing Y_{i1} are imputed from their regression on observed X_i and Y_{i2}, while the missing Y_{i2} are imputed from their regression on observed X_i and Y_{i1}. Similarly, extensions to higher dimensional Y_i are straightforward (e.g., see Problems 34–37). Rubin (1986) illustrates the use of an implicit model to create multiple imputations in this context.

Problems Arising with Nonmonotone Patterns

Nonmonotone missingness complicates the modeling task, the estimation task, and the imputation task. The problems can be illustrated with bivariate Y_i where some units have Y_{i1} observed, some units have Y_{i2} observed, and some units have both Y_{i1} and Y_{i2} observed. Suppose the modeling task tentatively posits that Y_{i1} given X_i has a normal linear regression on univariate X_i, and that Y_{i2} given Y_{i1} and X_i has a normal linear regression on X_i, Y_{i1}, Y_{i1}^2, and $X_i Y_{i1}$, as suggested by theory or some preliminary regression analysis using the units with both Y_{i1} and Y_{i2} observed.

BEYOND MONOTONE MISSINGNESS

First note that the modeling task is more difficult because standard diagnostics are not immediately appropriate. Even when nonresponse is ignorable, diagnostic information for each regression involves all units in a not entirely transparent manner, whereas with monotone missingness, each univariate factor provides its own relevant diagnostic information. That is, if $Y_{[1]}$ were at least as observed as $Y_{[2]}$, the units with $Y_{[2]}$ observed would provide the diagnostic information about the regression of $Y_{[2]}$ on $(X, Y_{[1]})$ in the standard complete-data manner, and the units with $Y_{[1]}$ observed would provide the diagnostic information about the regression of $Y_{[1]}$ on X, also in the standard complete-data manner.

The estimation task is more difficult even when the parameters of the two regressions are *a priori* independent because it does not reduce to two independent complete-data estimation tasks. Even large sample likelihood methods using the EM algorithm (Dempster, Laird, and Rubin, 1977) require much work to implement in the case described with nonlinear terms involving Y_{i1} because of difficulties in the E step for the units with Y_{i1} missing; that is, the conditional distribution of Y_{i1} given X_i and Y_{i2} for units with Y_{i1} missing must be found, and this is not easy because of the dependence of the Y_{i2} regression on nonlinear terms in Y_{i1}.

The imputation task is more difficult even with known values for the parameter, also because of the need to find conditional distributions not explicitly formulated in the modeling task, such as the distribution of Y_{i1} given X_i and Y_{i2} just discussed in the context of the E step of EM.

Although these problems are stated in terms of explicit models, analogous issues arise when using implicit models for imputation with nonmonotone patterns. More work is needed to develop good tools for the case of nonmonotone missingness. Currently, there exist five general solutions: (1) discard data to create a monotone pattern, (2) assume conditional independence among blocks of variables to create independent monotone patterns, (3) use an explicit, analytically tractable but possibly not fully appropriate model and related approximations, (4) iteratively apply methods for monotone patterns with explicit models, and (5) use the SIR (sampling/importance resampling) algorithm under appropriate explicit models.

Discarding Data to Obtain a Monotone Pattern

Because of the problems inherent when faced with nonmonotone patterns and the relative simplicity when patterns are monotone, an obvious procedure is to discard Y_{ij} values that destroy monotone missingness; this is especially attractive if only a few Y_{ij} values need to be discarded. As a specific example, Table 5.6, abstracted from Marini, Olsen, and Rubin

TABLE 5.6. Example from Marini, Olsen and Rubin (1980) illustrating how to obtain a monotone pattern of missing data by discarding data; 1 = observed, 0 = missing.

		Y Variables			Number	Percentage
Pattern	X	Block 1	Block 2	Block 3	of Cases	of Cases
A	1	1	1	1	1594	36.6
B	1	1^a	1^a	0	648	14.9
C	1	1	0	1^b	722	16.6
D	1	1^a	0	0	469	10.8
E	1	0	0	1^b	499	11.5
F	1	0	0	0	420	9.6
					4352	100.0

^{a} Observations falling outside monotone pattern 2 (X more observed than block 3; block 3 more observed than block 1; block 1 more observed than block 2).
^{b} Observations falling outside monotone pattern 1 (X more observed than block 1; block 1 more observed than block 2; block 2 more observed than block 3).

(1980), displays a pattern of missingness in three blocks of Y variables, where the footnotes a and b refer to choices of which values to discard in order to create a monotone pattern. Of course, an important practical issue is which values to discard with minimum loss of information, and this depends not only on the number of data points but also the interdependence among the variables. For instance, in Table 5.6, if the last block of Y variables were perfectly predictable from X, discarding according to footnote b would result in no loss of information. This is a virtually unstudied problem, although posed in Rubin (1974). Similarly, if only one missing Y_{ij} destroys a monotone pattern, an obvious approach is to fill it in using some naive method and proceed using careful models assuming that value were known. Strategies for doing this well are again virtually unstudied.

Assuming Conditional Independence Among Blocks of Variables to Create Independent Monotone Patterns

A second general approach is to take advantage of presumed structure in the relationships among the variables to create several independent monotone patterns. The idea is readily explained by a simple example. Suppose in Table 5.6 that the 499 observations in block 3 pattern E were missing instead of observed. Further suppose that for the purposes of imputation, we are willing to accept the modeling assumption that block 2 and block 3 variables are conditionally independent given block 1 variables and X.

BEYOND MONOTONE MISSINGNESS

Then for purposes of imputation, we have returned to the monotone case: missing block 1 values are imputed from the modeling of block 1 variables given X, missing block 2 values are imputed from the modeling of block 2 variables given block 1 variables and X, and missing block 3 values are imputed from the modeling of block 3 variables given block 1 variables and X. In some cases, such assumptions of conditional independence may be perfectly reasonable, as when block 1 variables and X are extensive background measurements, and block 2 and block 3 consist of variables that prior information suggests are basically unrelated, especially given X. Of course, analyses of multiply-imputed data sets created under this independence assumption will tend to confirm this independence even if it is not an accurate reflection of reality. Consequently, a multiply-imputed data set should include indications of any such assumptions used to create the imputations.

Using Computationally Convenient Explicit Models

A third general approach when faced with nonmonotone missingness is to forego finely tuned univariate models and instead use an algorithmically convenient explicit model for multivariate Y_i, create imputations under this model, and then try to fix up any resulting inconsistencies by editing checks. Perhaps the most obvious convenient model is the usual multivariate normal linear model. For this model, the EM algorithm (Orchard and Woodbury, 1972; Beale and Little, 1975; Dempster, Laird, and Rubin, 1977; Little and Rubin, 1987) can be used to find maximum-likelihood estimates, and those estimates combined with the second derivative matrix of the log likelihood at the maximum can be used to define an approximate normal posterior distribution for the parameters, thereby defining the estimation task. With drawn values of the parameters, the simple sweep operator (Beaton, 1964; Dempster, 1969; Little and Rubin, 1987) can be used to find all required normal distributions from which imputations are easily made.

With discrete multivariate Y_i, which define a contingency table, again the EM algorithm can be used to find maximum-likelihood estimates (Hartley, 1958; Dempster, Laird, and Rubin, 1977; Little and Rubin, 1987), and large sample normal approximations based on this estimate and the second derivative of the log likelihood at the maximum can be used to specify the estimation task. The imputation task consists of finding conditional probabilities in various cells of the contingency table given observed margins.

Recent work by Little and Schluchter (1985) (also see Little and Rubin, 1987) extends the EM algorithm to cases where multivariate Y_i involves both normal and discrete variables. Developing realistic and algorithmically convenient models in the presence of nonmonotone missing data is important. Nevertheless, these methods tend to suffer from the fact that (a) the models are not highly tuned to specific data sets, as can be relatively easily done with models in the case of monotone missingness, and (b) the large-sample approximations to the posterior distributions of parameters can be inadequate in practice.

Iteratively Using Methods for Monotone Patterns

Some recent work suggests the possibility of applying methods for monotone patterns iteratively in order to handle cases with nonmonotone patterns. In a sense, these methods can be viewed as extensions and combinations of the ideas of the EM algorithm and multiple imputation: at the E step, instead of finding the conditional expectation of the complete-data log posterior, multiply impute from the current estimate of the posterior distribution; at the M step, instead of finding the value of θ that maximizes the complete-data likelihood, represent the current posterior distribution as a mixture of complete-data posterior distributions using the multiple imputations. The ideas were independently proposed with different emphases and implementations by Tanner and Wong (1987), as a technique to simulate the posterior distribution of θ, and Li (1985), as a method for drawing imputations. The methods are easily implemented in many particular cases but are still not completely general. Nevertheless, the techniques and their extensions seem to be very promising and worthy of further development.

The Sampling / Importance Resampling Algorithm

Within the context of multiply-imputed public-use data bases, both the fraction of values missing and the number of imputations per missing value must be modest for the size of the auxiliary matrix of imputations to be reasonable. An algorithm for creating draws from the posterior distribution of Y_{mis} that is designed to operate in this restricted environment of modest m and modest fractions of missing information is the sampling/importance resampling (SIR) algorithm (Rubin, 1987). This algorithm is noniterative and can be applied even when the missingness-modeling structure is such that the imputation task is intractable, both attractive features with public-use data bases, which are often quite large with complex patterns of missing values.

The SIR algorithm first requires a good approximation to the joint posterior distribution of (Y_{mis}, θ), say

$$\tilde{\Pr}(Y_{mis}, \theta | X, Y_{obs}) = \tilde{\Pr}(\theta | X, Y_{obs}) \tilde{\Pr}(Y_{mis} | X, Y_{obs}, \theta),$$

which is positive for all possible (Y_{mis}, θ) at the observed (X, Y_{obs}), and, second, needs to be able to calculate the *importance ratios*

$$r(Y_{mis}, \theta) \propto \Pr(Y|X, \theta)\Pr(\theta)/\tilde{\Pr}(Y_{mis}, \theta|X, Y_{obs})$$

for all possible (Y_{mis}, θ) at the observed (X, Y_{obs}). When the imputation task is tractable,

$$\tilde{\Pr}(Y_{mis}|X, Y_{obs}, \theta) = \Pr(Y_{mis}|X, Y_{obs}, \theta),$$

and only $\Pr(\theta|X, Y_{obs})$ has to be approximated; also, then the importance ratios do not depend on Y_{mis}:

$$r(Y_{mis}, \theta) \propto \Pr(Y_{obs}|X, \theta)\Pr(\theta)/\tilde{\Pr}(\theta|X, Y_{obs}).$$

Some Details of SIR

SIR has three basic steps:

1. Draw M values of (Y_{mis}, θ) from the approximation $\tilde{\Pr}(Y_{mis}, \theta|X, Y_{obs})$, where M is large relative to m.
2. Calculate the importance ratios $r(Y_{mis}, \theta)$ for each drawn value of (Y_{mis}, θ), say $r_1, \ldots, r_M$.
3. Draw m values of Y_{mis} with probability proportional to $r_1, \ldots, r_M$ from the M values drawn in step 1. Methods for such drawing appear in the survey literature on pps sampling, for example, Cochran (1977, Chapter 9).

If the imputation task is tractable, only values of θ need be drawn in step 1 and used to calculate the importance ratios in step 2. In step 3, then m values of θ are drawn and the corresponding m values of Y_{mis} are drawn from the tractable $\Pr(Y_{mis}|X, Y_{obs}, \theta)$.

The rationale for the SIR algorithm is based on the fact that as $M/m \to \infty$, the m values of (Y_{mis}, θ) drawn in step 3 are drawn with probabilities given by

$$\frac{\tilde{\Pr}(Y_{mis}, \theta|X, Y_{obs})r(Y_{mis}, \theta)}{\iint \tilde{\Pr}(Y_{mis}, \theta|X, Y_{obs})r(Y_{mis}, \theta)\, dY_{mis}\, d\theta},$$

which equals $\Pr(Y_{mis}, \theta|X, Y_{obs})$ after straightforward algebraic manipulation. The choice of an adequate ratio M/m for practice depends on the fraction of missing information, γ. Smaller γ implies satisfactory inferences from smaller M/m for two reasons: first, $\tilde{\Pr}(Y_{mis}, \theta | X, Y_{obs})$ should be a more accurate approximation to $\Pr(Y_{mis}, \theta | X, Y_{obs})$, and second, final inferences from a multiply-imputed data set will be less sensitive to the imputed values. In common practice, $M/m = 20$ may be more than adequate, especially considering that with modest m, accurately approximating the tails of $\Pr(Y_{mis} | X, Y_{obs})$ is of limited importance. Also, in many situations it may make sense to choose M/m adaptively. For example, initially set $M/m = 2$; in step 2 calculate the variance of $\log(r_1), \ldots, \log(r_m)$ to measure the adequacy of the approximate posterior distribution; and then select the actual ratio M/m as a function of this variance. The development of adaptive SIR algorithms seems to warrant further study.

Example 5.7. An Illustrative Application of SIR

The following artificial example is taken from Rubin (1987) and illustrates the application of SIR to a problem that cannot be directly addressed by EM, the Li (1985), or the Tanner and Wong (1987) algorithms. Suppose with bivariate Y and no X the proposed model has Y_i i.i.d. with

$$Y_{i1} | \theta \sim N(\mu_1, \sigma_1^2) \qquad (5.6.1)$$

$$Y_{i2} | Y_{i1}, \theta \sim N(\alpha + \beta Y_{i1} + \gamma Y_{i1}^2, \tau^2), \qquad (5.6.2)$$

where

$$\theta = (\mu_1, \log \sigma_1, \alpha, \beta, \gamma, \log \tau) \qquad (5.6.3)$$

and

$$\Pr(\theta) \propto \text{const.} \qquad (5.6.4)$$

In the sample of n units, n_1 have only Y_{i1} observed, n_2 have only Y_{i2} observed, and n_{12} have both Y_{i1} and Y_{i2} observed, where $n_{12} > n_1 + n_2 > n_2 > n_1$. The missingness is not monotone.

The only specific issue in implementing the SIR algorithm here is choosing good initial approximations for $\Pr(\theta | X, Y_{obs})$ and $\Pr(Y_{mis} | X, Y_{obs}, \theta)$. The numerators of the importance ratios $r(Y_{mis}, \theta)$ are easily evaluated since the density $\Pr(Y | X, \theta) \Pr(\theta)$ is simply the product over the n units of the two normal densities (5.6.1) and (5.6.2).

Regarding $\Pr(\theta|X, Y_{obs})$, it can be easily approximated by independent normal densities:

$$\log \sigma_1 | X, Y_{obs} \sim N\left(\log s_1, [2(n_1 + n_{12} - 1)]^{-1}\right), \quad (5.6.5)$$

$$\mu_1 | X, Y_{obs} \sim N\left(\bar{y}_1, s_1^2/(n_1 + n_{12})\right), \quad (5.6.6)$$

$$\log \tau | X, Y_{obs} \sim N\left(\log \hat{\tau}, [2(n_{12} - 3)]^{-1}\right), \quad (5.6.7)$$

and

$$(\alpha, \beta, \gamma) | X, Y_{obs} \sim N\left((\hat{\alpha}, \hat{\beta}, \hat{\gamma}), \hat{\tau}^2 C\right), \quad (5.6.8)$$

where $\bar{y}_1$ and s_1^2 are the mean and variance of the $(n_1 + n_{12})$ observations of Y_{i1}, and $(\hat{\alpha}, \hat{\beta}, \hat{\gamma}, \hat{\tau}^2, C)$ are the standard least-squares summaries obtained by regressing Y_{i2} on $(1, Y_{i1}, Y_{i1}^2)$ using the n_{12} observations of (Y_{i1}, Y_{i2}). Similarly, $\Pr(Y_{mis}|X, Y_{obs})$ can be easily approximated by $n_1 + n_2$ independent normal densities:

$$n_1 \text{ missing } Y_{i2}|X, Y_{obs}, \theta \sim N\left(\hat{\alpha} + \hat{\beta} Y_{i1} + \hat{\gamma} Y_{i1}^2, \hat{\tau}^2\right), \quad (5.6.9)$$

and

$$n_2 \text{ missing } Y_{i1}|X, Y_{obs}, \theta \sim N\left(a + b Y_{i2} + c Y_{i2}^2, s_e^2\right), \quad (5.6.10)$$

where (a, b, c, s_e^2) are the standard least-squares summaries obtained by regressing Y_{i1} on $(1, Y_{i2}, Y_{i2}^2)$ using the n_{12} observation of (Y_{i1}, Y_{i2}). Thus, $\Pr(Y_{mis}, \theta|X, Y_{obs})$ is the product of the $(n_1 + n_2 + 3)$ univariate normal densities specified by (5.6.5), (5.6.6), (5.6.7), (5.6.9), and (5.6.10), and the trivariate normal density specified by (5.6.8); it is therefore easy to draw from at step 1, and easy to evaluate as the denominator of the importance ratios in step 2. Better approximations are available, especially for small n_{12}, but it is not clear whether it is worth the effort to develop them in the context of SIR relative to increasing the ratio M/m.

PROBLEMS

1. Describe how each of the examples of surveys suffering from nonresponse described in Chapter 1 are more complicated than the case of univariate Y_i, no X_i, simple random sampling, and ignorable nonresponse.

2. Why might it be reasonable to assume ignorable nonresponse when creating one set of multiple imputations even when the reasons for nonresponse are not fully understood?

3. If Y_i is univariate and there are many respondents at each distinct value of X_i, which multiple imputation methods would you recommend under various conditions (e.g., Y_i continuous versus dichotomous)? What are the reasons for your recommendations? Hint: check the examples in Chapter 4.

4. Describe the changes in the Census Bureau's hot-deck procedures and comment on the Lillard, Smith, and Welch (1986) criticisms of these procedures, also see Rubin (1980d, 1983d).

5. Describe Statistics Canada's imputation methods and how they could be modified to create multiple imputations. Hint: check the NAS volumes on incomplete data.

6. Briefly review the observational study literature on matching methods, and suggest which methods might be most relevant to survey practice.

7. How could "matching on predicted Y_i" be defined with multivariate Y_i, for example, using canonical variates?

8. Describe advantages and disadvantages of imputing by matching on the best prediction of missing values.

9. Consider other ways of describing the tasks needed to create imputed values. What are advantages and disadvantages of the modeling–estimation–imputation description?

10. Classify the methods for imputation discussed in Madow, Nisselson, and Olkin (1983).

11. What is the meaning and source of the expressions "cold-deck imputation," and "hot-deck imputation," "donor," and "donee"? Hint: check the NAS volumes.

12. What is the logic underlying a sequential rather than a random draw hot-deck, and is there evidence of systematic differences in practice (e.g., East to West versus West to East)? Hint: check the NAS volumes.

13. Define weighting cell estimators and adjustment cells, and describe the reasons for their names. Hint: Check Little and Rubin (1987).

14. Describe kernel-density estimation imputation methods, such as proposed by Sedransk and Titterington (1984).

15. Review literature on nonresponse published in recent issues of the *Proceedings of the American Statistical Association Section on Survey Research Methods*.

PROBLEMS

16. Compare and contrast two general methods for reducing multivariate X_i to univariate X_i: by creating regression scores (i.e., predicted Y_i), and by creating propensity scores (i.e., predicted probability of being a nonrespondent).

17. Describe a standard analysis in which the objective is to predict Y from X yet $\theta_{Y|X}$ and θ_X are not *a priori* independent. Hint: Suppose Y_i is dichotomous and X_i is continuous. Comment on the "backwards" modeling, referring perhaps to Efron (1975) or Rubin (1984).

18. Is Result 5.3 true if $\theta_{Y|X}$ and θ_X are not *a priori* independent? Provide an example to illustrate essential ideas.

19. Describe the logic behind the modifications offered in Example 5.2 and defend the standardization employed. Compare analytically or by simulation the various methods that have been defined in simple but revealing cases. Is Santos (1981) relevant to this case?

20. Compare the balanced repeated replication (BRR) method of handling nonresponse, in which each half sample is imputed independently, with multiple imputation. For example, how many effective degrees of freedom does each method yield for standard errors? What is the effect of deviations from modeling assumptions? Assuming BRR would be used with complete data, how would it be used with multiple imputation, and are there then any advantages to multiple imputation over single imputation?

21. Describe the estimation and imputation tasks when Y_i is modeled as approximately proportional to X_i as in Example 2.4. Describe the extension needed when $\log(Y_i)$ is modeled as approximately proportional to X_i.

22. Describe an approach to modeling symmetric long-tailed Y_i given X_i and the associated estimation and imputation tasks. Hint: suppose Y_i given X_i is a linearly translated t distribution on g (known) degrees of freedom, and examine Dempster, Laird, and Rubin (1980).

23. Suppose X_i indicates K strata, where Y_i is normal within each stratum, and the number of observations of Y_i in each stratum is modest. Posit an exchangeable normal prior distribution for the stratum parameters $\{(\mu_k, \sigma_k^2), k = 1, \ldots, K)\}$ and describe the estimation and imputation tasks. Defend your choice of distribution for $\tau^2 =$ the variance of the μ_k. Hint: see Dempster, Rubin, and Tsutakawa (1981), Rubin (1981b), and Morris (1983a).

24. Relate the model in Problem 23 to the literature on James-Stein (1961) estimation, empirical-Bayes estimation (e.g., Efron and Morris, 1975),

and Bayesian hyperparameter estimation (e.g., Lindley and Smith, 1972).

25. Comment on the two versions of smoothing illustrated by the regression and hyperparameter approaches; can they be described to appear nearly identical in a formal sense, and why or why not?

26. Suppose Y_i is multivariate but only one component can ever be missing. Is the pattern monotone? Describe the tasks of a multiple-imputation procedure in this case.

27. Suppose Y_i is multivariate and either fully observed or fully missing [i.e., $R_i = (1,\ldots,1)$ or $(0,\ldots,0)$]. Is this a monotone pattern? Assuming Y_i given X_i and θ is multivariate normal, describe two multiple-imputation procedures, the first based on repeated univariate procedures, the second based on a multivariate procedure, which imputes the entire missing vector at once. Describe advantages and disadvantages of the approaches.

28. Describe why Result 5.4 needs the independence in (5.4.2) and Result 5.5 does not.

29. Describe the estimation and imputation tasks when bivariate Y_i has a monotone-independent structure with positive Y_{i1} such that $\sqrt{Y_{i1}}$ given X_i is approximately normal, and dichotomous Y_{i2} given Y_{i1} and X_i is a logistic regression. Hint: Watch out for negative Y_{i1}; see Rubin (1983b).

30. Are the summaries involving posterior means and variances in Section 5.5 appropriate for dichotomous Y_{i1} and lognormal Y_{i2}? What sort of summaries might be more informative? Also, show how to calculate the fractions of missing information given there.

31. Criticize the study described in Section 5.5 and design an improved version.

32. Suppose (Y_i, X_i) is bivariate normal and the estimands are the regression parameters of X_i on Y_i, $(\alpha, \beta, \log \sigma)$, where Y_i is more likely to be missing for larger X_i. Design a simulation study comparing various inferential methods such as (i) discard missing units, (ii) best value imputation, (iii) hot-deck imputation, and so on. Vary the correlation between Y_i and X_i, but not the other parameters (why not?). Vary the sample size and response rate. Compare multiple and simple imputation procedures with respect to accuracy, efficiency, and validity of interval estimation.

33. In the setup of Problem 32, why aren't "errors-of-measurement in predictor variables" a problem when estimating β using multiple imputation? Assume large samples and numbers of imputations to

PROBLEMS

show that in the case of Problem 32, using repeated imputations under an explicit normal model yields valid inferences.

34. Using the sweep operator, describe how to impute Y_{mis} in Example 5.6 extended to include multivariate X_i. Hint: see Rubin and Thayer (1978).

35. Describe how to apply techniques for univariate Y_i to the following case with file-matching missingness: $(Y_{i1}, \ldots, Y_{ik})$ is fully observed for n_1 units, $(Y_{i,k+1}, \ldots, Y_{ip})$ is fully observed for n_2 units where $n = n_1 + n_2$ so that the two sets of Y variables are never jointly observed. Be explicit for the multivariate normal model.

36. Extend the results of Problem 35 to the case with the missingness in $(Y_{i1}, \ldots, Y_{ik})$ monotone for the n_1 units and the missingness in $(Y_{i,k+1}, \ldots, Y_{ip})$ monotone for the n_2 units.

37. Extend the results of Problem 36 to more than two blocks of never jointly observed variables.

38. Summarize why nonmonotone patterns of missingness create special problems.

39. Describe advantages and disadvantages of (a) discarding data to create a monotone pattern versus (b) using an analytically tractable but not fully appropriate model.

40. Outline an approach to nonmonotone missingness that produces imputed values only for a small subset of the missing values by using a convenient model, and proceeds by capitalizing on the resulting monotone pattern created by having observed values for this subset.

41. Consider procedures for deciding how to lose the minimum amount of information when creating a monotone missingness by discarding data, for example, under a normal model.

42. Describe advantages and disadvantages of the following method for imputing data with a nonmonotone pattern. Fit a multivariate normal by EM; use the estimated parameters to measure the information lost by discarding data; discard data to create monotone missingness with minimal loss of information; assume a monotone-independent structure and create multiple imputations using a sequence of univariate (possibly nonnormal) modeling, estimation, and imputation tasks.

43. Give an example showing that not discarding data with the Census Bureau's hot-deck can create inconsistencies with variables not used to define the matching.

44. Outline in some detail the estimation and imputation tasks with a general pattern of missingness using EM with the p-variate normal.

Comment on similarities between the E step of EM and the imputation task.

45. Repeat Problem 44 for the p-way contingency table model and the combined normal linear/contingency table model.
46. Comment on the following discussion:

    There are two kinds of errors when using a simple hot-deck multiple-imputation procedure:

    1. Between-imputation variability is underestimated because the parameters do not vary across imputations.
    2. Within-imputation variability is overestimated because matches are never exact with respect to observed variables.

    In practice, these two errors will tend to trade off evenly, yielding properly estimated variability. Moreover, the m imputations for each missing value are no longer ordered, and so many more completed data sets can be created, thereby increasing efficiency of estimation and dramatically increasing the degrees of freedom available to estimate the between-imputation component of variance.

47. Outline a study to address any unresolved issues arising from Problem 46.
48. Relate the iterative methods proposed in Li (1985) and Tanner and Wong (1987).
49. Describe cases where the methods in Problem 48 work easily and other cases when they do not (e.g., consider the multivariate normal and then the discussion following Example 5.4 regarding difficulties with the E step).
50. Outline a general plan for handling nonmonotone missingness using the SIR algorithm.
51. Review the literature on pps sampling.
52. Show why SIR works when $M/m \to \infty$.
53. Develop an expression for the actual probabilities of selection using SIR when M/m is finite.
54. Using the result in Problem 53, formulate advice on the choice of M/m.
55. Develop specific guidance for the use of an adaptive version of SIR.

PROBLEMS

56. Show that the fraction of missing information in Example 5.7 is modest when nonresponse is unconfounded and provide explicit expressions for γ.

57. For confounded but ignorable nonresponse in Example 5.7, is the fraction of missing information still modest? What is the maximum fraction of missing information under these conditions?

58. Write down the exact posterior distribution for (Y_{mis}, θ) in Example 5.7 assuming $n_2 = 0$.

CHAPTER 6

Procedures with Nonignorable Nonresponse

6.1. INTRODUCTION

When nonresponse is nonignorable, there exists response bias in the sense that a respondent and nonrespondent with exactly the same values of variables observed for both have systematically different values of variables missing for the nonrespondent. Since there are no data to estimate this nonresponse bias directly, inferences will necessarily be sensitive to assumptions about similarities between respondents and nonrespondents. These assumptions can be addressed by scientific understanding or related data from other surveys.

Displaying Sensitivity to Models for Nonresponse

Commonly, such external sources of information will not be precise enough to establish accurately the ways in which respondents differ from nonrespondents. Consequently, an important component of a satisfactory analysis when nonignorable nonresponse is being considered is the display of sensitivity to different plausible specifications for the response mechanism. Within the context of multiple imputation, this display of sensitivity is accomplished by creating two or more imputations under each of several models for nonresponse, creating separate inferences under each model, and comparing the conclusions reached under each of the models.

INTRODUCTION

The Need to Use Easily Communicated Models

Since commonly no one model will be obviously more realistic than some others, and users of the survey will have to make judgments regarding the relative merits of the various inferences reached under different nonresponse models, easily communicated models should be used whenever possible. Mathematically complicated models may not only be difficult to apply, but they may also be difficult to communicate to users of surveys. Even though sophisticated models are potentially more accurate reflections of reality, in common practice it may be wise to use simply described modifications of more standard ignorable models, such as the "20% increase over ignorable value" model used to illustrate ideas in Section 1.6.

Transformations to Create Nonignorable Imputed Values from Ignorable Imputed Values

Transformations more complicated than

$$(\text{nonignorable imputed } Y_i) = 1.2 \times (\text{ignorable imputed } Y_i) \quad (6.1.1)$$

could just as easily have been used in the example of Section 1.6 to illustrate a simple method for creating systematic differences between respondents and nonrespondents with the same value of X_i. For example, consider the transformation

$$(\text{nonignorable imputed } Y_i) = \exp[a + b \times \log(\text{ignorable } Y_i)],$$

which changes the location, scale, and shape of the ignorable imputed values depending on the constants a and b.

The advantages of such fixed transformations of univariate Y_i are their easy creation and communication. Extension of this technique to the case of multivariate Y_i is straightforward when the pattern of missing data is monotone, since in this case all imputations can be created using a sequence of univariate models. Some formal details with explicit models are presented in Section 6.3, and the idea can easily be used with implicit models as well as explicit ones.

Other Simple Methods for Creating Nonignorable Imputed Values Using Ignorable Imputation Models

There exists a variety of simple methods other than fixed transformations that can be used to distort ignorable imputation methods into producing nonignorable imputations. For example, only a fixed percentage of the imputed values might be distorted by a fixed transformation; specifically, for a random 50% of the imputed values, apply the transformation (6.1.1). Such a distortion might be appropriate when there exists a suspicion that there are varying reasons for nonresponse, with only some of the nonrespondents being different from respondents.

Another possibility is to distort the probability of drawing values to impute using a function of the value to be imputed. For example, for each value to be imputed, draw 10 values under an ignorable model and then choose the value to impute by drawing from the 10 with probability proportional to each value. Such a distortion might be appropriate when the probability of nonresponse on Y_i is thought to be related to the value of Y_i, as when probability of nonresponse on income questions is thought to be proportional to $Y_i = \log(\text{income})$ even after adjustment for covariates, X.

Essential Statistical Issues and Outline of Chapter

The modeling, estimation, and imputation tasks with nonignorable nonresponse can all be technically demanding in realistic settings, and it is easy to lose sight of the essential statistical issues regarding sensitivity of inference to assumptions that are not directly verifiable and the implications of the assumptions in such cases. That is, to impute $Y_{i,mis}$, the missing values for the ith unit, we need to estimate the conditional distribution of $Y_{i,mis}$ given the observed values $(X_i, Y_{i,obs}, R_{i,inc})$. But if we examine all units with the same value of R as the ith unit, and try to estimate the conditional distribution of $Y_{i,mis}$ given $(X_i, Y_{i,obs})$ from these units, say using a regression model, we will fail because by the definitions of $R_{i,inc}$ and $Y_{i,mis}$, none of these units has any observed values of $Y_{i,mis}$.

With ignorable nonresponse, the conditioning on the value of R_{inc} is irrelevant, and so standard models and methods can be employed, which estimate the conditional distribution of $Y_{i,mis}$ given $(X_i, Y_{i,obs})$ using the other units who have the components in $Y_{i,mis}$ observed. But with nonignorable nonresponse, special models are needed. These issues are clearly exposed in the simple case of univariate Y_i, no X_i, and the usual i.i.d. modeling structure, and so this case is the topic of Section 6.2.

Section 6.3 considers the formal tasks that are required to create nonignorable imputed values and identifies two distinct ways of performing the modeling task, which can lead to quite different estimation and imputation tasks in practice. Section 6.4 illustrates the first, the mixture modeling approach, in the context of Example 1.4 involving missing school principals' responses. Section 6.5 illustrates the second, the selection modeling approach, in the context of Example 1.2 with nonreported income data in the Current Population Survey (CPS).

When nonignorable nonresponse is considered likely, good survey design often tries to pursue some nonrespondents to obtain responses. Data from followed-up nonrespondents can be extremely useful for creating realistic imputed values for those nonrespondents without follow-up data. Section 6.6 considers general issues concerning multiple imputation with followed-up nonrespondents and shows how multiple imputation can lead to standard inferences in this case. Section 6.7 illustrates the mixture modeling approach to surveys with followed-up nonrespondents in the context of Example 1.4 concerning missing responses to drinking behavior questions. Throughout, we assume that the sampling mechanism is ignorable and focus on the issues created by a nonignorable response mechanism.

6.2. NONIGNORABLE NONRESPONSE WITH UNIVARIATE Y_i AND NO X_i

Consider the case with univariate Y_i, no X_i, an ignorable sampling mechanism, but a nonignorable response mechanism. By equation (2.6.3), inferences require a specification for $\Pr(Y, R)$, which following de Finetti's theorem, can be written as

$$\Pr(Y, R) = \int \prod_{i=1}^{N} f_{YR}(Y_i, R_i|\theta)\Pr(\theta)\,d\theta. \tag{6.2.1}$$

The Modeling Task

In this form, the modeling task must specify the distribution $f_{YR}(Y_i, R_i|\theta)$ and the prior distribution $\Pr(\theta)$. Without loss of generality, we can write

$$f_{YR}(Y_i, R_i|\theta) = f_{Y|R}(Y_i|R_i, \theta)f_R(R_i|\theta). \tag{6.2.2}$$

Again without loss of generality, we can write the factors on the right-hand side of (6.2.2) as

$$f_{Y|R}(Y_i|R_i, \theta) = \begin{cases} f_0(Y_i|\theta_0), & \text{if } R_i = 0 \\ f_1(Y_i|\theta_1), & \text{if } R_i = 1 \end{cases} \tag{6.2.3}$$

and

$$f_R(R_i|\theta) = \theta_R^{R_i}(1 - \theta_R)^{1-R_i} \tag{6.2.4}$$

where θ_0, θ_1, and θ_R are functions of θ: θ_0 the parameter of the data for the nonrespondents, θ_1 the parameter of the data for the respondents, and θ_R the parameter giving the probability of being a respondent. In general θ_0, θ_1, and θ_R are *a priori* dependent.

The Imputation Task

To impute missing values from the Bayesian perspective, we need the posterior distribution of the missing values, Y_{mis}:

$$\Pr(Y_{mis}|Y_{obs}, R_{inc}) = \int \Pr(Y_{mis}|Y_{obs}, R_{inc}, \theta) \Pr(\theta|Y_{obs}, R_{inc}) \, d\theta.$$

From (6.2.1)–(6.2.3).

$$\Pr(Y_{mis}|Y_{obs}, R_{inc}, \theta) = \prod_{mis} f_0(Y_i|\theta_0),$$

so that

$$\Pr(Y_{mis}|Y_{obs}, R_{inc}) = \int \prod_{mis} f_0(Y_i|\theta_0) \Pr(\theta_0|Y_{obs}, R_{inc}) \, d\theta_0. \quad (6.2.5)$$

Equation (6.2.5) implies that to impute Y_{mis}, first draw a value of θ_0 from its posterior distribution, say θ_0^*, and then draw the components of Y_{mis} as i.i.d. from $f_0(Y_i|\theta_0^*)$. Thus, the imputation task requires only the posterior distribution of θ_0.

The Estimation Task

Writing

$$\Pr(\theta) = \Pr(\theta_0|\theta_1, \theta_R) \Pr(\theta_1, \theta_R)$$

where $\Pr(\theta_0|\theta_1, \theta_R)$ is the prior distribution of θ_0 given (θ_1, θ_R), it is simple to see from (6.2.1)–(6.2.4) that

$$\Pr(\theta_0|Y_{obs}, R_{inc}) = \int \Pr(\theta_0|\theta_1, \theta_R) \Pr(\theta_1, \theta_R|Y_{obs}, R_{inc}) \, d(\theta_1, \theta_R) \quad (6.2.6)$$

where $\Pr(\theta_1, \theta_R|Y_{obs}, R_{inc})$ is the posterior distribution of (θ_1, θ_R):

$$\Pr(\theta_1, \theta_R|Y_{obs}, R_{inc}) \propto \prod_{obs} f_1(Y_i|\theta_1) \prod_{inc} \theta_R^{R_i}(1-\theta_R)^{1-R_i} \Pr(\theta_1, \theta_R). \quad (6.2.7)$$

NONIGNORABLE NONRESPONSE WITH UNIVARIATE Y_i 207

Equation (6.2.6) implies that in order to draw a value of θ_0 from its posterior distribution, first draw a value of (θ_1, θ_R) from its posterior distribution (6.2.7), say (θ_1^*, θ_R^*), and then draw θ_0 from its conditional prior (= posterior) distribution given $(\theta_1, \theta_R) = (\theta_1^*, \theta_R^*)$. This fact formally demonstrates the sensitivity of inference to untestable assumptions: the drawn value of θ_0 is entirely dependent on its *prior* distribution given (θ_1, θ_R). Thus with nonignorable nonresponse, we can learn about the distribution for nonrespondents, governed by θ_0, only via untestable prior ties to the distribution for respondents, governed by θ_1, or in some unusual cases using prior ties to the distribution of response itself, governed by θ_R.

Two Basic Approaches to the Modeling Task

One basic approach to the modeling task is to follow (6.2.2)–(6.2.4) and specify: (a) the distribution of the population data for respondents governed by θ_1, (b) the distribution of the data for nonrespondents governed by θ_0, and (c) the distribution of response governed by the probability of response, θ_R, where (θ_1, θ_0) is *a priori* independent of θ_R. This approach has been called the mixture modeling approach since it defines the population distribution of Y as a mixture of respondents and nonrespondents. The second basic approach is to specify $f_{YR}(Y_i, R_i|\theta)$ in (6.2.1) by writing

$$f_{YR}(Y_i, R_i|\theta) = f_Y(Y_i|\theta_Y) f_{R|Y}(R_i|Y_i, \theta_{R|Y})$$

and specifying: (a) the distribution of the population data governed by θ_Y and (b) the distribution of the response mechanism governed by $\theta_{R|Y}$, where θ_Y and $\theta_{R|Y}$ are *a priori* independent. This approach has been called the selection modeling approach because of the specification for the response mechanism that selects units to be respondents.

Example 6.1. The Simple Normal Mixture Model

A very simple example of the mixture modeling approach specifies $f_1(Y_i|\theta_1)$ as $N(\mu_1, \sigma_1^2)$ and $f_0(Y_i|\theta_0)$ as $N(\mu_0, \sigma_0^2)$, where $\theta_1 = (\mu_1, \log \sigma_1)$, $\theta_0 = (\mu_0, \log \sigma_0)$. With respect to the prior distribution of θ, θ_R is *a priori* independent of (θ_1, θ_0) with $\Pr(\theta_R) \propto$ const on $(0,1)$, $\Pr(\theta_1) \propto$ const, and $\Pr(\theta_0|\theta_1)$ specifies that *a priori*, $\mu_0 = k_\mu + \mu_1$ and $\sigma_0 = k_\sigma \sigma_1$ where k_μ and k_σ are fixed constants. Thus the parameters for nonrespondents are a fixed transformation of the parameters for respondents. Letting $\bar{y}_1$ and s_1^2 be the sample mean and variance of the n_1 respondents' values, one set of imputed values for Y_{mis} is created as follows:

1. Draw a value of σ_1^2 from its posterior distribution, say $\sigma_{1*}^2 = s_1^2(n_1 - 1)/x$ where x is a χ^2 random variable on $n_1 - 1$ degrees of freedom.
2. Draw a value of μ_1 from its posterior distribution given $\sigma_1^2 = \sigma_{1*}^2$, say $\mu_{1*} = \bar{y}_1 + \sigma_{1*}z$ where z is a $N(0, 1)$ deviate.
3. Draw a value of (μ_0, σ_0^2) from its posterior distribution given $\mu_1 = \mu_{1*}$ and $\sigma_1^2 = \sigma_{1*}^2$, say

$$\mu_{0*} = k_\mu + \mu_{1*}$$

and

$$\sigma_{0*} = k_\sigma \sigma_{1*}.$$

4. Draw the $n_0 = n - n_1$ values of Y_{mis} as i.i.d. $N(\mu_{0*}, \sigma_{0*}^2)$, or equivalently, for $i \in mis$,

$$Y_i = k_\mu + \mu_{1*} + k_\sigma \sigma_{1*} z_i, \quad z_i \stackrel{\text{i.i.d.}}{\sim} N(0, 1).$$

A similar but not identical specification alters only $f_0(Y_i|\theta_0)$: *a priori* $\theta_0 = \theta_1$ but

$$f_0(Y_i|\theta_0) = f_1\left(\left.\frac{Y_i - k_\mu}{k_\sigma}\right|\theta_1\right).$$

Under this fixed-transformation nonignorable model, which has

(nonignorable imputed value) = $k_\mu + k_\sigma$ (ignorable imputed value),

after drawing a value of θ_1 from its posterior distribution, say $(\mu_{1*}, \log \sigma_{1*})$, the n_0 missing values of $(Y_i - k_\mu)/k_\sigma$ are drawn as i.i.d. $N(\mu_{1*}, \sigma_{1*}^2)$, or equivalently, for $i \in mis$

$$Y_i = k_\mu + k_\sigma(\mu_{1*} + \sigma_{1*} z_i).$$

Such fixed transformations of imputed values were introduced in Section 1.6.

As these two specifications illustrate, the mixture-modeling approach easily generates nonignorable values using (i) standard ignorable models and (ii) prior descriptions of distortions to parameters or imputed values. The mixture modeling approach is thus very well suited to the exploration of sensitivity to various nonignorable models.

Example 6.2. The Simple Normal Selection Model

Suppose it is known that in the population the Y_i are nearly normally distributed and that nonresponse becomes increasingly likely with larger Y_i. For instance, perhaps $Y_i = \log(\text{income})$ and individuals with larger incomes are more likely to be nonrespondents on income questions. One model that has been proposed in cases like this is to specify $f_Y(Y_i|\theta_Y)$ as

$$(Y_i|\theta_Y) \sim N(\mu, \sigma^2)$$

and to specify $f_{R|Y}(R_i|Y_i, \theta_{R|Y})$ as

$$\Pr(R_i = 1|Y_i, \theta_{R|Y}) = g(\gamma_0 + \gamma_1 Y_i)$$

where γ_0 and γ_1 are parameters to be estimated and $g(\cdot)$ is some cumulative distribution function. For instance, if $g(\cdot)$ is the normal cumulative distribution function, then $f_{R|Y}(R_i|Y_i, \theta_{R|Y})$ can be specified via an unobserved variable U_i such that $R_i = 1$ if $U_i > \varepsilon$ and $= 0$ if $U_i < \varepsilon$, where U_i given θ is $N(0, 1)$ and has correlation ρ with Y_i; here $\theta = (\mu, \log \sigma, \rho, \varepsilon)$ with all parameters mutually independent. Models like this have been studied by among others, Heckman (1976) and Nelson (1977), and are called normal selection models because they append to the standard normal model a model that selects which values of Y_i will be observed.

The proportion of Y_i that are observed $(\Sigma_{inc} R_i/n)$, addresses the estimation of ε; for example, with 50% response rate, the obvious point estimate of ε is 0 whereas with 97.5% response rate, the obvious estimate of ε is 1.96. Since given θ the n values of Y_i are i.i.d. $N(\mu, \sigma^2)$, the skewness in the n_1 observed values of Y_i addresses the estimation of ρ. For example, suppose the response rate is 50%; if the observed values of Y_i look exactly normally distributed, an obvious estimate of ρ is 0, whereas if the observed values of Y_i look exactly like the left half of a normal distribution, an obvious large-sample point estimate of ρ is -1. Since the observed skewness addresses the value of ρ and thus the extent of nonresponse bias, the results of this model are extremely sensitive to the symmetry assumption of the normal specification. Finally, if we knew values for ε and ρ, we could estimate μ and σ^2; for example, if $\varepsilon = 0$ and $\rho = -1$, then the maximum observed Y_i is an obvious estimate of μ, whereas if $\rho = 0$, $\bar{y}_1$ and s_1^2 are the obvious estimates of μ and σ^2.

Although it is clear that under the model specifications there exists information about the parameters θ, the posterior distribution of θ does not have a neat form for standard prior distributions and so the estimation task is not straightforward. There are various iterative algorithms available to find maximum-likelihood estimates and to evaluate the second derivatives of the log likelihood at the maximum used for standard large-sample normal approximations, but there is little study of the accuracy of such approximations in realistic cases.

Furthermore, although the distribution of Y_i given θ is simple, as is the distribution of R_i given Y_i and θ, the distribution that is needed for the imputation task is that of Y_i given $R_i = 0$ and θ, which is a stochastically truncated normal distribution. A simple, but possibly inefficient, procedure is to (i) draw Y_i from its distribution given θ, say Y_i^*, (ii) draw R_i from its conditional distribution given $Y_i = Y_i^*$ and θ, say R_i^*, (iii) impute the drawn value of Y_i if $R_i^* = 0$ (i.e., missing), and (iv) return to step (i) if $R_i^* = 1$.

A selection model that includes X variables and *a priori* constraints on regression coefficients is used to generate multiple imputations of missing log(income) in Section 6.4.

6.3. FORMAL TASKS WITH NONIGNORABLE NONRESPONSE

The general modeling, estimation, and imputation tasks with nonignorable nonresponse are extensions and complications of the corresponding tasks with ignorable nonresponse. Consequently, the structure and results presented here closely parallel those of Chapter 5.

The Modeling Task—Notation

Since the sampling mechanism is ignorable, results in Section 2.6 imply that the posterior distribution of unobserved Y values follow from specifications for $\Pr(Y|X)$ and the response mechanism, $\Pr(R|X, Y)$, or equivalently, from the joint distribution of (X, Y, R), $\Pr(X, Y, R)$. By the usual appeal to de Finetti's theorem, the rows of (X, Y, R) can be modeled as i.i.d. given some parameter θ:

$$\Pr(X, Y, R) = \int \Pr(X, Y, R|\theta)\Pr(\theta) = \int \prod_{i=1}^{N} f_{XYR}(X_i, Y_i, R_i|\theta)\Pr(\theta)\, d\theta.$$

(6.3.1)

Since X is fully observed, it is convenient to write

$$f_{XYR}(X_i, Y_i, R_i|\theta) = f_{YR|X}(Y_i, R_i|X_i, \theta_{YR|X}) f_X(X_i|\theta_X) \quad (6.3.2)$$

where $\theta_{YR|X}$ and θ_X are functions of θ. Commonly, but not necessarily, $\theta_{YR|X}$ and θ_X are specified to be *a priori* independent:

$$\Pr(\theta) = \Pr(\theta_{YR|X})\Pr(\theta_X). \quad (6.3.3)$$

Furthermore, since by definition R is more observed than Y, it is also convenient to write:

$$f_{YR|X}(Y_i, R_i|X_i, \theta_{YR|X}) = f_{Y|XR}(Y_i|X_i, R_i, \theta_{Y|XR}) f_{R|X}(R_i|X_i, \theta_{R|X})$$

$$(6.3.4)$$

where $\theta_{Y|XR}$ and $\theta_{R|X}$ are functions of $\theta_{YR|X}$ and thus of θ. In some cases, $\theta_{Y|XR}$ is specified to be *a priori* independent of $(\theta_{R|X}, \theta_X)$:

$$\Pr(\theta) = \Pr(\theta_{Y|XR})\Pr(\theta_{R|X}, \theta_X). \quad (6.3.5)$$

Two General Approaches to the Modeling Task

As illustrated in Section 6.2 for the case of univariate Y_i and no X_i, two basic approaches to modeling nonignorable nonresponse can be identified. One, the mixture modeling approach, accepts (6.3.4) and (6.3.5), and seems better suited to the straightforward investigation of sensitivity of inferences using multiple imputation. The other approach, the selection modeling approach, follows naturally from considering the additional specification needed to modify the ignorable modeling task and has

$$f_{YR|X}(Y_i, R_i|X_i, \theta_{YR|X}) = f_{Y|X}(Y_i|X_i, \theta_{Y|X}) f_{R|YX}(R_i|X_i, Y_i, \theta_{R|XY}) \quad (6.3.6)$$

where $\theta_{Y|X}$ is *a priori* independent of $(\theta_{R|XY}, \theta_X)$,

$$\Pr(\theta) = \Pr(\theta_{Y|X})\Pr(\theta_{R|XY}, \theta_X). \quad (6.3.7)$$

Similarities with Ignorable Case

The general modeling structure in (6.3.1)–(6.3.3) is formally similar to that in (5.2.1) and (5.2.2) with ignorable nonresponse, where now (Y_i, R_i) plays the role of Y_i, and $\theta_{YR|X}$ plays the role of $\theta_{Y|X}$. Hence, some results in Chapter 5 on the imputation and estimation tasks formally apply with this notational change. Other results in Chapter 5 apply with only minor modification. Henceforth, we assume the structure in (6.3.1) and use the notation in (6.3.2), (6.3.3), and that previously established in Chapter 5 (e.g., $Y_{i,mis}$ = the components of Y missing for the ith unit).

The Imputation Task

The objective is to draw Y_{mis} from its posterior distribution, which can be written as

$$\Pr(Y_{mis}|X, Y_{obs}, R_{inc}) = \int \Pr(Y_{mis}|X, Y_{obs}, R_{inc}, \theta)\Pr(\theta|X, Y_{obs}, R_{inc})\, d\theta.$$

As with ignorable nonresponse, our plan is to draw a value of θ from its posterior distribution, $\Pr(\theta|X, Y_{obs}, R_{inc})$, say θ^*, and then draw a value of Y_{mis} from its conditional posterior distribution given the drawn value of θ^*, $\Pr(Y_{mis}|X, Y_{obs}, R_{inc}, \theta = \theta^*)$. Repeating this process m times creates m draws from the joint posterior distribution of (Y_{mis}, θ). Ignoring the drawn values of θ gives m draws from the posterior distribution of Y_{mis}. The reason for drawing values of both θ and Y_{mis} is that under the i.i.d. model in (6.3.1), the task of drawing Y_{mis} given θ is relatively simple.

Result 6.1. The Imputation Task with Nonignorable Nonresponse

Given θ, the $Y_{i,mis}$ are *a posteriori* independent with distribution depending on θ only through $\theta_{YR|X}$:

$$\Pr(Y_{mis}|X, Y_{obs}, R_{inc}, \theta) = \prod_{ms} \Pr(Y_{i,mis}|X_i, Y_{i,obs}, R_{i,inc}, \theta_{YR|X}).$$

The result follows from application of the same arguments as in Result 5.1 after substituting (Y, R) for Y and $\theta_{YR|X}$ for $\theta_{Y|X}$, and then integrating over R_{exc}.

Result 6.2. The Imputation Task with Nonignorable Nonresponse When Each Unit Is Either Included in or Excluded from the Survey

Suppose $I_i = (1,\ldots,1)$ or $(0,\ldots,0)$ for all i; then given θ, the $Y_{i,mis}$ are *a posteriori* independent with distribution depending on θ only through $\theta_{Y|XR}$, which is a function of $\theta_{YR|X}$:

$$\Pr(Y_{mis}|X, Y_{obs}, R_{inc}, \theta) = \prod_{ms} \Pr(Y_{i,mis}|X_i, Y_{i,obs}, R_i, \theta_{Y|XR}).$$

Result 6.2 follows by the same argument as Result 6.1 with the observation that when $I_i = (1,\ldots,1)$ or $(0,\ldots,0)$, for $i \in ms$ R_i is fully observed

FORMAL TASKS WITH NONIGNORABLE NONRESPONSE   213

so that the distribution of $Y_{i,mis}$ given $(X_i, Y_{i,obs}, R_{i,inc})$ and $\theta_{YR|X}$ follows from $f_{Y|XR}(Y_i|X_i, R_i, \theta_{Y|XR})$.

The Estimation Task

Result 6.1 shows that the only function of θ that is needed for the imputation task is $\theta_{YR|X}$, the parameter of $f_{YR|X}(Y_i, R_i|X_i, \theta_{YR|X})$ specified in the modeling task. Result 6.2 shows that in the common case when each unit is either included or excluded from the survey, the only function of θ that is needed is $\theta_{Y|XR}$, the parameter of $f_{Y|XR}(Y_i|X_i, R_i, \theta_{Y|XR})$. Thus, the objective of the estimation task is to compute the posterior distribution of $\theta_{YR|X}$, or more simply, $\theta_{Y|RX}$ in some cases, so that a random draw of the parameter can be made. Generally, this task is quite demanding even with apparently simple models. In analogy with the case of ignorable nonresponse, some simplification can occur by specifying $\theta_{YR|X}$ to be *a priori* independent of θ_X. Furthermore, additional simplification can occur by specifying $\theta_{Y|XR}$ to be *a priori* independent of $(\theta_X, \theta_{R|X})$. Notwithstanding these simplifications, however, it is not always wise to specify such *a priori* independence, especially with *a priori* knowledge to the contrary. For example, if (Y, X) is known to have nearly a common correlation between all pairs of variables, obviously the observed correlations among the X variables help to estimate the correlations involving the Y variables.

Result 6.3. The Estimation Task with Nonignorable Nonresponse When $\theta_{YR|X}$ Is *a Priori* Independent of θ_X

Suppose $\theta_{YR|X}$ and θ_X are *a priori* independent as in (6.3.3). Then they are *a posteriori* independent; moreover, the posterior distribution of $\theta_{YR|X}$ involves only (a) the specifications $f_{YR|X}(\cdot|\cdot)$ and $\Pr(\theta_{YR|X})$ and (b) data from units included in the survey.

Result 6.3 is immediate from Result 5.2 after substituting (Y, R) for Y and $\theta_{YR|X}$ for $\theta_{Y|X}$.

Result 6.4. The Estimation Task with Nonignorable Nonresponse When $\theta_{Y|XR}$ Is *a Priori* Independent of $(\theta_{R|X}, \theta_X)$ and Each Unit Is Either Included in or Excluded from the Survey

Suppose $I_i = (1, \ldots, 1)$ or $(0, \ldots, 0)$, and $\theta_{Y|XR}$ and $(\theta_{R|X}, \theta_Y)$ are *a priori* independent as in (6.3.5). Then they are *a posteriori* independent; moreover, the posterior distribution of $\theta_{Y|XR}$ involves only (a) the specifications $f_{Y|XR}(\cdot|\cdot)$ and $\Pr(\theta_{Y|XR})$ and (b) data from units with some Y_i observed.

Result 6.4 follows from Result 5.2 or Result 6.3 noting that for units included in the survey, R can be treated as part of X since it is a fully observed variable.

Result 6.5. The Imputation and Estimation Tasks with Nonignorable Nonresponse and Univariate Y_i

Suppose Y_i is univariate; then given $\theta_{Y|XR}$ the $Y_{i,\,mis}$ are *a posteriori* independent; furthermore, if $\theta_{Y|XR}$ and $(\theta_{R|X}, \theta_X)$ are *a priori* independent, they are *a posteriori* independent, and the posterior distribution of $\theta_{Y|XR}$ involves only (a) the specifications $f_{Y|XR}(\cdot|\cdot)$ and $\Pr(\theta_{Y|XR})$ and (b) data from respondents.

This result follows immediately from Results 6.1 and 6.4 since if Y_i is univariate each unit is automatically either included in or excluded from the survey. It is stated as a separate result because of the practical importance of the case with univariate Y_i.

Monotone Missingness

Suppose the pattern of missingness is monotone as defined in Section 5.5, and assume for notational simplicity that $I_i = (1, \ldots, 1)$ or $I_i = (0, \ldots, 0)$. We saw in Chapter 5 that with ignorable nonresponse, by choosing a distinct modeling structure that corresponds to the missingness pattern, the modeling, estimation, and imputation tasks could be reduced to a sequence of independent univariate modeling, estimation, and imputation tasks. There are several possible extensions to the case of nonignorable nonresponse, but the most obvious is within the mixture modeling perspective where R is simply treated like a fully observed X variable. Following the notation in (5.4.1)–(5.4.4), the jth factor in the distinct monotone structure, f_{ij}, is now the conditional distribution of Y_{ij} given $(X_i, Y_{i1}, \ldots, Y_{i,j-1})$ and R_i, and this distribution's parameter, θ_j. It is usually convenient to model the nonignorability of Y_{ij}, that is the dependence of this distribution on R_i, only through R_{ij}, because then the estimation and imputation tasks are formally identical to a sequence of p independent univariate nonignorable estimation and imputation tasks. The notation required to be fully precise is more cumbersome than insightful, and reinterpreting the notation established in Chapter 5 conveys the essential ideas. The conclusions are summarized in Result 6.6 without proof.

Result 6.6. The Estimation and Imputation Tasks with a Monotone-Distinct Structure and a Mixture Model for Nonignorable Nonresponse

Suppose $\theta_{Y|XR}$ and $(\theta_{R|X}, \theta_X)$ are *a priori* independent, and suppose that the missingness pattern is monotone with $Y_{[1]}$ at least as observed as $Y_{[2]}$, and so on, as defined by (5.4.3). Furthermore, factorize the densities $f_{Y|XR}(Y_i|X_i, R_i, \theta_{Y|XR})$ and $\Pr(\theta_{Y|XR})$ as in (5.4.1) and (5.4.2), respectively, where the conditioning on R_i is implicit in (5.4.1) and $\theta = (\theta_1, \ldots, \theta_p)$ is interpreted as $\theta_{Y|XR}$.

Then using this notation and the shorthand f_{ij} in (5.4.4), we have that the θ_j are *a posteriori* independent with distributions given by (5.4.5), for $j = 1, \ldots, p$; (5.4.5) is formally identical to the posterior distribution of the parameter of the distribution of univariate Y_{ij} given $(X_i, Y_{i1}, \ldots, Y_{i,j-1}, R_i)$ based on the n_j units with $(X_i, Y_{i1}, \ldots, Y_{i,j-1}, R_i)$ fully observed. If this distribution depends on R_i only through R_{ij}, then this posterior distribution is formally identical to that obtained in an estimation task for univariate Y_i using mixture modeling of nonignorable nonresponse; moreover, the imputation task for Y_i is then also formally identical to that with univariate Y_i.

Selection Modeling and Monotone Missingness

Corresponding results for selection modeling require far more restrictive assumptions. Specifically, the estimation and imputation tasks reduce to a sequence of univariate estimation and imputation tasks when using selection models with data having a monotone pattern of missingness, if the following conditions hold:

1. $\theta_Y, \theta_{R|XY}$, and θ_X are *a priori* mutually independent (this simply defines selection modeling).
2. Given $(X_i, Y_i, \theta_{R|XY})$, the $R_{i1}, \ldots, R_{ip}$ are mutually independent with parameters that are *a priori* independent, and the distribution of R_{ij} depends on Y_i only through Y_{ij}:

$$f_{R|XY}(R_i | X_i, Y_i, \theta_{R|XY}) = \prod_{j=1}^{p} f_{R|XY}^{[j]}(R_{ij} | X_i, Y_{ij}, \theta_{R|XY}^{[j]})$$

where

$$\Pr(\theta_{R|XY}) = \prod_{j=1}^{p} \Pr(\theta_{R|XY}^{[j]}).$$

6.4. ILLUSTRATING MIXTURE MODELING USING EDUCATIONAL TESTING SERVICE DATA

The data we use to illustrate the mixture modeling approach to nonresponse comes from the survey of high school principals introduced here as Example 1.1. The analysis presented here represents the first attempt (1972) to use some of the basic ideas presented in this text. In fact, the original effort did not perform any imputations, but used analytic methods exclusively. Nevertheless, in the published article presenting the example (Rubin, 1977a), the answers are described as if they might have been obtained via multiple imputation: "One can think of the method given here as simply summarizing the results of simulations, where one uses the respondents to generate 'reasonable' hypothetical responses for the nonrespondents." The aim was to display sensitivity of the sample mean to prior assumptions about the similarity between respondents and nonrespondents. The posterior distribution of the complete-data sample mean, $\bar{y}$, was derived analytically under a family of normal linear regression mixture models. The posterior distribution of the complete-data standard deviation, s, was not considered, so that complete-data inferences were not compared. That is, the posterior distributions of $\bar{y}$ rather than of $\overline{Y}$ were compared under different prior assumptions. Although the results in Rubin (1977a) were derived analytically, we will describe them as if they were found by the simulation techniques we have presented here, and extend those results to include complete-data inferences for $\overline{Y}$.

The Data Base

In 1971, 660 high school principals were asked to fill out a compensatory reading questionnaire consisting of 80 dependent variables. The schools in the survey were chosen from a very large population, consisting of most high schools in the United States, using 35 background variables, X (e.g., median income and racial composition in ZIP code area; enrollment, number of rooms, and approximate percentage reading below grade level). The survey design used X with an unconfounded probability sampling mechanism to choose the 660 schools so that the mean of X in the sample was close to the mean of X in the population.

Of the 660 principals included in the survey, 188 refused to answer the questionnaire. Consequently, all 80 dependent variables are missing for all 188 nonrespondent schools. For simplicity, we focus on one important dependent variable, "17B", how often compensatory reading was conducted during regular school hours in time released from other work, coded 0, .2, .5, .8, 1.0 indicating the proportion of time released. This variable, Y, recorded for the 472 respondent schools, coupled with the 35 covariates, X, recorded for all schools, comprise the data base for this example.

The Modeling Task

The mixture model we posit for studying nonignorable nonresponse specifies $f_{Y|XR}(Y_i|X_i, R_i, \theta_{Y|XR})$ by a normal linear regression for respondents' Y_i given X_i,

$$(Y_i|X_i, R_i = 1, \theta_{Y|XR}) \sim N(\alpha_1 + X_i\beta_1, \sigma_1^2),$$

and for nonrespondents' Y_i given X_i,

$$(Y_i|X_i, R_i = 0, \theta_{Y|XR}) \sim N(\alpha_0 + X_i\beta_0, \sigma_0^2),$$

where $\theta_{Y|XR} = (\alpha_1, \beta_1, \log \sigma_1, \alpha_0, \beta_0, \log \sigma_0)$ is *a priori* independent of both $\theta_{R|X}$, the parameter governing the probability of being a respondent at each value of X, and θ_X, the parameter of the marginal distribution of X.

The modeling task is completed by specifying a prior distribution for $\theta_{Y|XR}$. Since the number of respondents is relatively large, for simplicity the standard noninformative prior distribution was used for the respondents' parameters, $\Pr(\alpha_1, \beta_1, \log \sigma_1) \propto \text{const}$. The crucial prior specification for investigating sensitivity of inference with nonignorable nonresponse is the conditional distribution of the nonrespondents' parameters $(\alpha_0, \beta_0, \log \sigma_0)$ given the respondents' parameters. For simplicity we set $\sigma_0 = \sigma_1 = \sigma$. The prior distribution of (α_0, β_0) given $(\alpha_1, \beta_1, \sigma)$ is the product of two independent distributions. First, β_0 is normally distributed about β_1,

$$(\beta_0|\alpha_1, \beta_1, \log \sigma) \sim N(\beta_1, c_\beta^2 \beta_1 \beta_1'),$$

where c_β is a nonnegative constant specifying the *a priori* similarity of β_0 and β_1. The corresponding prior distribution of α_0 is specified indirectly and treats the mean X for respondents, $\overline{X}_1$, as fixed (more formally, the prior distribution of $\theta_{Y|RX}$ described here is implicitly conditional on X). Specifically, the average Y value at $\overline{X}_1$ for nonrespondents in the population, $\eta_0 = (\alpha_0 + \overline{X}_1\beta_0)$, is normally distributed about the average Y value for respondents in the population, $\eta_1 = (\alpha_1 + \overline{X}_1\beta_1)$,

$$(\eta_0|\alpha_1, \beta_1, \sigma^2) \sim N(\eta_1, c_\eta^2 \eta_1^2),$$

where c_η is a nonnegative constant specifying the *a priori* similarity of η_0 and η_1.

Clarification of Prior Distribution Relating Nonrespondent and Respondent Parameters

If $c_\beta = c_\eta = 0$, the prior specifications imply an ignorable nonresponse mechanism; for positive c_β or c_η, the response mechanism is nonignorable. The first constant, c_β, determines how similar the slopes of Y on X are likely to be for respondents and nonrespondents, whereas c_η determines how similar the expected value of Y is likely to be for groups of respondents and nonrespondents with X means equal to the respondent X mean in the population.

The parameter c_β is the prior coefficient of variation (the standard deviation divided by the mean) for the nonrespondents' regression coefficients. That is, letting $\beta_1^{(j)}$ and $\beta_0^{(j)}$ be the jth components of β_1 and β_0, respectively, the investigator is 95% sure that $\beta_0^{(j)}$ will fall in the interval

$$\beta_1^{(j)}(1 \pm 1.96c_\beta).$$

Note we assume that c_β is the same for all X variables and that there is no *a priori* bias (i.e., the interval for β_0 is centered at β_1). Of course, more complicated assumptions are possible. We use coefficients of variation because they seem well suited to summarize subjective notions of large and small in this example (e.g., $c_\beta = 1\%$ seems small, whereas $c_\beta = 100\%$ seems large).

Even if both (a) the regression coefficients (slopes) of Y on X and (b) the distributions of X were identical for respondents and nonrespondents, the two groups might have different Y means; c_η reflects this difference as the coefficient of variation for the mean Y in a group of nonrespondents whose X distribution is the same as the X distribution for the respondents. Since we are assuming a linear relation between Y and X, we really only need consider a group of nonrespondents whose X mean is $\bar{X}_1$. Thus, since η_1 is the expected value of Y for a group of respondents whose average X is $\bar{X}_1$, the investigator is 95% sure that the expected Y for a group of nonrespondents whose average X is $\bar{X}_1$ lies in the interval

$$\eta_1(1 \pm 1.96c_\eta).$$

Again, note we assume that there is no *a priori* bias.

Comments on Assumptions

The rationale for this particular parameterization is primarily that it proved conceptually useful to the psychologists and educators for whom the survey was taken. They were able to use other related data and their experience to conjecture about c_β and c_η, and they felt that the formulation facilitated easy communication. Also, the primary reason for using the normal linear model is that it is easy to handle formally; it should be reasonable for these data because Y_i is a five-category ordered response. For some dependent variables having mostly zero or one responses, a log-linear model would probably be more appropriate.

In other problems (or even in this problem with other types of researchers), other models and parameterizations might be more useful. An obvious restriction to relax is the prior covariance of β_0 being proportional to $\beta_1\beta_1^t$. For example, different coefficients of variation could be used for different background variables and the prior covariance matrix could be diagonal instead of rank one.

No matter how we parameterize a nonresponse problem, however, two points are crucial. Both follow from the fact that if the propensity to be a nonrespondent is some probabilistic function solely of the recorded background variables X, then the response mechanism is ignorable and the conditional distribution of Y given X is the same for nonrespondents and respondents (in our model, $c_\beta = c_\eta = 0$). Hence, the observed distributions of X for respondents and nonrespondents should not influence the researcher's choice of c_β and c_η, except perhaps when used in conjunction with other information. Furthermore, as more background variables are recorded, the investigator should be more confident that if the Y means for respondents and nonrespondents differ, the difference should be reflected in the observed distribution of the background variables; in our example, both c_β and c_η should decrease as more background variables are recorded. This second point follows because as more background variables are recorded, it becomes more likely (to be precise, not less likely) that the propensity to be a nonrespondent is some probabilistic function of the recorded background variables.

The Estimation Task

The estimation task is very easy for the model just specified. First, the posterior distribution of (η_0, β_0) given $(\eta_1, \beta_1, \sigma^2)$ is the same as its prior distribution given $(\eta_1, \beta_1, \sigma^2)$, specified previously. Second, the posterior distribution of (η_1, β_1) given σ^2 is identical in form to the usual sampling distribution of regression coefficients in normal linear models; that is, the posterior distribution of β_1 given σ^2 is normal with mean equal to the least-squares coefficient of the Y_i on X_i for respondents, $\hat{\beta}_1 = V\left[\Sigma_{obs}(X_i - \bar{X}_1)^t Y_i\right]$, and covariance $\sigma^2 V$ where $V = \left[\Sigma_{obs}(X_i - \bar{X}_1)^t(X_i - \bar{X}_1)\right]^{-1}$—for simplicity we assume $\bar{X}_1 = \Sigma_{obs} X_i/n_1 = \Sigma_1^N X_i/N$, and the posterior distribution of η_1 given σ^2 and β_1 is normal with mean $\bar{y}_1$ and variance $\sigma^2/472$. Third, the posterior distribution of σ^2 is $436 s_1^2$ times an inverted χ^2 on $472 - 35 - 1 = 436$ degrees of freedom, where s_1^2 is the residual mean square from the respondents' regression.

The Imputation Task

The imputation task is also easy. First draw from the posterior distribution of respondents' parameters, $(\eta_1, \beta_1, \log \sigma)$; then draw from the conditional posterior distribution of nonrespondents' parameters, (η_0, β_0), given respondents' parameters; and finally draw from the posterior distribution of the missing values, Y_{mis}, for nonrespondents. Repeat the process m times using independent random numbers to create m repeated multiple imputations of Y_{mis}.

Specifically, the following steps are followed with c_β and c_η fixed by the investigator:

1. Draw a value of σ^2 from its posterior distribution, say

$$\sigma_*^2 = 436 \times s_1^2/\chi_{436}^2$$

where χ_{436}^2 is a χ^2 random variable on 436 degrees of freedom.

2. Draw a value of β_1 from its conditional posterior distribution given the drawn value of σ^2, say

$$\beta_{1*} = \hat{\beta}_1 + \sigma_*[V]^{1/2}Z_1$$

where $[V]^{1/2}$ is a square root of V, such as the triangular square root obtained by a Cholesky factorization, and Z_1 is a vector of 35 i.i.d. $N(0,1)$ deviates.

3. Draw a value of η_1 from its conditional posterior distribution given the drawn values of σ^2 and β_1, say

$$\eta_{1*} = \bar{y}_1 + (\sigma_*^2/472)z_1$$

where z_1 is an independent $N(0,1)$ deviate.

4. Draw a value of β_0 from its conditional posterior distribution given the drawn values of σ^2, β_1, and η_1, say

$$\beta_{0*} = \beta_{1*} + c_\beta \beta_{1*} Z_0$$

where Z_0 is an independent $N(0,1)$ deviate.

5. Draw a value of η_0 from its conditional posterior distribution given the drawn values of σ^2, β_1, η_1, and β_0, say

$$\eta_{0*} = \eta_{1*} + c_\eta \eta_{1*} z_0$$

where z_0 is an independent $N(0,1)$ deviate.

6. Finally, draw a value of Y_{mis} from its posterior distribution given the drawn value of $\theta_{Y|XR}$; that is, for each of the 188 nonrespondents let

$$Y_{i*} = \eta_{0*} + (X_i - \bar{X}_1)\beta_{0*} + \sigma_* z_i$$

where the z_i are new i.i.d. $N(0,1)$ deviates.

Analysis of Multiply-Imputed Data

Variable 17B, *time released from other class work*, was considered particularly important because if time for compensatory reading was released from other class work, rather than given after school hours or during free study periods, some educators felt that it might indicate a serious need for more resources in the school. The mean of 17B for respondents was .39 and its standard deviation was .40.

Table 6.1 provides summary statistics and repeated-imputation 95% intervals for $\bar{Y}$ under each of eight imputation models, one with ignorable nonresponse ($c_\beta = c_\eta = 0$), and seven with nonignorable nonresponse ($c_\beta > 0$ or $c_\eta > 0$). For each set of repeated imputations under one model, the estimate based on the completed data set was the complete-data regression-adjusted estimate of $\bar{Y}$, and its associated variance was the standard complete-data variance (see Cochran, 1977, Chapter 7, and Problem 14 of Chapter 2). Because we are assuming the sample and population means of X are essentially the same, the complete-data estimate equals the mean of the 660 Y_i values, $\bar{y}$, with associated complete-data variance $s^2/660$, where s^2 is the residual mean square in the regression of Y on X using the 660 units in the sample. The resultant infinite m repeated-imputation interval is $E_*(\bar{y}) \pm 1.96\,[E_*(s^2/n) + V_*(\bar{y})]^{1/2}$ where $E_*( )$ and $V_*( )$

TABLE 6.1. Summary of repeated–imputation intervals for variable 17B in educational example.[a]

Model		Expectations over Repeated Imputations			95% Interval for $\bar{Y} \times 10^2$	Percentage Information Missing Due to Nonresponse
c_β	c_η	$E_*(\bar{y}) \times 10^2$	$V_*(\bar{y}) \times 10^4$	$E_*(s^2/n) \times 10^4$		
0.0	0.0	31	1.21	2.18	(27, 35)	36
0.0	0.1	31	2.47	3.22	(26, 36)	43
0.0	0.2	31	6.24	6.25	(24, 38)	50
0.0	0.4	31	21.69	18.16	(18, 44)	54
0.4	0.0	31	1.39	2.32	(27, 35)	37
0.4	0.1	31	2.66	3.41	(26, 36)	44
0.4	0.2	31	6.43	6.80	(24, 38)	49
0.4	0.4	31	21.87	19.40	(18, 44)	53

[a] In fact, because the raw data for this study were no longer available, the values in this table combine values from analytic results published in Rubin (1977a) and approximations based on some summary statistics available from earlier draft reports.

refer to the mean and variance over an infinite number of imputations. The center of all the intervals is .31, which differs from the mean Y for respondents, $\bar{y}_1$, because the distribution of the background variables differs for respondents and nonrespondents, but is the same for all values of (c_β, c_η) because the model for the response mechanism has (β_0, η_0) *a priori* centered at (β_1, η_1). Although it might be of interest to study models with *a priori* bias, Table 6.1 simply indicates the increased uncertainty that results when nonignorable models are considered. The fractions of information missing due to nonresponse are greater than the nonresponse rate (28%) because respondents and nonrespondents have different distributions of X.

First note that the value of c_β is not critical, whereas the value of c_η is. That is, the 95% interval widths are insensitive to the value of c_β but quite sensitive to the value of c_η. Recall that c_η is the prior coefficient of variation for the mean Y for a group of nonrespondents with the same distribution of X as the respondents. Because many relevant background variables are being controlled, it might be reasonable to assume that groups of nonrespondents and respondents with the same X means would have similar Y means. Thus it might be reasonable to suppose c_η to be modest. But the insensitivity to values of c_β indicates that Y is not that well predicted by X. Moreover, even for modest c_η, the interval for $\bar{Y}$ can easily be twice as large as when assuming ignorable nonresponse, and consequently, before sharp inferences are drawn assuming ignorable nonresponse, nonresponse bias due to different Y means at the same value of X should be carefully considered. The type of nonresponse bias resulting from different regression coefficients of Y on X for respondents and nonrespondents does not seem to be important. Finally, as expected intuitively, the percentage of information missing due to nonresponse becomes larger as the nonignorability becomes more extreme.

6.5. ILLUSTRATING SELECTION MODELING USING CPS DATA

The data set we use to illustrate the selection modeling approach comes from the CPS–IRS–SSA exact match file, introduced in Section 5.5. Two important differences between this application by Greenlees, Reece, and Zieschang (1982a, b) (hereafter GRZ) and the one in Section 5.5, are that "total wages" (the IRS wage and salary variable) rather than "OASDI benefits" is the outcome variable, and the model being used is nonignorable rather than ignorable. The nonignorable model is a selection/stochastic-censoring model of the type illustrated in Example 6.2 applied to log(wages). The manner in which administrative data and nonresponse status are used, however, is the same as in Section 5.5: Nonresponse on the CPS total wage and salary question indicates which units are to be considered nonrespondents, but the wage and salary data come from IRS administrative records, rather than from the CPS wage and salary questions.

The Data Base

The data base thus consists of background variables, X, from the CPS; response indicator, R, from the CPS; and total wages, Y, from IRS records. As with the CPS example in Section 5.5, the data base is restricted, in this case to exclude households (units) whose head was unmarried, under 14, had farm or self-employment income, and so on; GRZ provide details. The result is a data set with 4954 CPS respondents, 410 nonrespondents due to refusal to answer the CPS wage and salary question, and 151 nonrespon-

TABLE 6.2. Background variables X for GRZ example on imputation of missing incomes.

Variable	A Priori Zero Coefficients in Prediction Equations	
	When Predicting Y, β	When Predicting R, δ
Constant		
Education		
White		
North		
South		
West		
Central city		0
Suburb		0
Education squared		0
Experience		0
Experience squared		0
Professional		0
Sales		0
Craft		0
Laborer		0
Construction		0
Manufacturing		0
Transportation		0
Trade		0
Service		0
Personal	0	
Age	0	

dents who did not answer the question for other reasons (e.g., lack of knowledge). GRZ perform analyses under both the narrow and broad definitions of nonresponse; thus the complete sample size is either 5364 or 5515 with either 410 or 561 nonrespondents, corresponding to nonresponse rates of either 8 or 10% depending on the definition of nonresponse.

The Modeling Task

The normal selection model used to model the data specifies that the outcome variable, $Y_i = \log(\text{wage})$, has a normal linear regression on X_i

$$(Y_i|X_i, \theta) \sim N(X_i\beta, \sigma^2)$$

where X_i includes variables indicating region of the country, education, experience, industry, and so on as given in the first column of Table 6.2; implicitly, GRZ assume the usual prior distribution on $(\beta, \log \sigma) \propto \text{const}$, except that two components of β are *a priori* zero, as indicated in the second column in Table 6.2. The probability of response is assumed to follow a logistic regression in predictors X_i and Y_i,

$$\Pr(R_i = 1|X_i, Y_i, \theta) = [1 + \exp(-X_i\delta - Y_i\gamma)]^{-1},$$

where *a priori* 14 components in the coefficient vector δ are zero as indicated in the third column of Table 6.2, and implicitly the prior distribution on (δ, γ) is proportional to a constant. The rationale for which coefficients in β and δ should be zero is rather mysterious to me, but this is common with selection models as applied to real data. *A priori*, $\theta_{Y|X} = (\beta, \log \sigma)$ and $\theta_{R|XY} = (\delta, \gamma)$ are both independent of θ_X. Because this specification depends on Y_i, it is nonignorable.

If a probit (normal probability) specification had been chosen for $\Pr(R_i = 1|X_i, Y_i, \theta)$ instead of logit, the model would have been identical to the one proposed by Nelson (1977). Lillard, Smith, and Welch (1986) applied the probit specification to CPS data, but did not use the exact match file; they also allowed the possibility that another transformation besides log was more appropriate.

Discussion of this approach appears in Little (1983c), Morris (1983b), and Rubin (1983d), where these selection models are criticized because their results can be terribly sensitive to the normality and *a priori* zero coefficient assumptions. GRZ are aware of this model sensitivity, and in fact are particularly interested in seeing how well the selection model works on the exact match file with its data from nonrespondents available to check the accuracy of the imputations.

The Estimation Task

GRZ approximate the posterior distribution of θ using the large-sample normal approximation, with mean equal to the maximum-likelihood estimate and variance–covariance matrix given by the inverse of the information matrix evaluated at the maximum. Since (Y_i, R_i) are observed for respondents and only R_i is observed for nonrespondents, the likelihood function can be written as

$$\prod_{obs} \Pr(Y_i, R_i = 1|\theta) \times \prod_{mis} \Pr(R_i = 0|\theta)$$

$$= \prod_{obs} \sigma^{-1}\phi((Y_i - X_i\beta)/\sigma)/[1 + \exp(-X_i\delta - Y_i\gamma)]$$

$$\times \prod_{mis} \int_{-\infty}^{\infty} \sigma^{-1}\phi((Y_i - X_i\beta)/\sigma)/[1 + \exp(-X_i\delta - Y_i\gamma)] \, dY_i \quad (6.5.1)$$

where $\phi(\cdot)$ is the standard normal density. Clearly, the estimation task is far more difficult for this selection model then the mixture model of Section 6.4. Furthermore, the quality of the results is to some extent dependent on the accuracy of the large-sample approximations.

GRZ maximized this likelihood using the generalized Gauss–Newton algorithm described in Berndt, Hall, Hall, and Hausman (1974). In fact, the likelihood function being maximized was slightly more complicated than the one given in equation (6.5.1): Since income is "topcoded" at $50,000, if the observed $Y_i = 50{,}000$ that respondent's factor in the likelihood is

$$\int_{\log(50{,}000)}^{\infty} \sigma^{-1}\phi((Y_i - X_i\beta)/\sigma)/[1 + \exp(-X_i\delta - Y_i\gamma)] \, dY_i.$$

GRZ also fit an ignorable model forcing $\gamma = 0$. Clearly, when $\gamma = 0$, the posterior distribution of (β, σ) is the standard one based on respondents' data alone. Tables in GRZ give estimated parameters and associated standard errors under both models and both definitions of nonresponse.

The Imputation Task

Ten values of missing total wages were imputed for each nonrespondent using this model with γ free to be estimated and also with γ fixed at zero, using both definitions of nonresponse. The imputation task was as follows:

1. Draw $\theta_{YR|X} = (\beta, \sigma, \gamma, \delta)$ from its approximating 30-variate normal posterior distribution; let $\theta^*_{YR|X} = (\beta_*, \sigma_*, \gamma_*, \delta_*)$ be the drawn value.

2. For $i \in mis$, draw (Y_i, R_i) from its joint posterior distribution given $\theta_{YR|X} = \theta^*_{YR|X}$:

   a. Let $Y_{i*} = X_i \beta_* + \sigma_* z_i$ where the z_i represent i.i.d. $N(0,1)$ deviates.

   b. Draw a uniform random number on $(0,1)$, u, and let $R_{i*} = 0$ if $u < \exp(-X_i \delta_* - Y_i \gamma_*)/[1 + \exp(-X_i \delta_* - Y_i \gamma_*)]$. Otherwise, let $R_{i*} = 1$.

   c. If $R_{i*} = 0$ (i.e., if $R_{i*} =$ the observed R_i), impute
   $$\min\{\hat{Y}_i, \log(50{,}000)\};$$
   otherwise return to step 2a.

This procedure generates one draw from the approximating posterior distribution of the missing data, Y_{mis}, and repeating it m times generates m repetitions from the posterior distribution of Y_{mis}. As with the estimation task, this imputation task is less direct than the imputation task for the mixture modeling example of Section 6.4.

Accuracy of Results for Single Imputation Methods

Table 6.3 gives the root-mean-squared errors (RMSE) of prediction when imputing the posterior mean log(wage) for each unit with θ fixed at its maximum-likelihood estimate. The errors are deviations of the imputed values from the true administrative values. Results are reported by columns for the nonignorable selection model and the ignorable version with γ a priori fixed at zero. For the row labeled "all nonrespondents," the nonignorable model was estimated from 5615 units, and the RMSE is for the 561 nonrespondents. For the row labeled "refusals," the nonignorable model was estimated from 5364 units, and the RMSE is for the 410 refusals. The ignorable model was estimated from the 4954 respondents, and the

TABLE 6.3. Root-mean-squared error of imputations of log-wage: Impute posterior mean[a] given θ fixed at MLE, $\hat{\theta}$.

	Ignorable Model	Nonignorable Model
All nonrespondents $N = 561$	0.486	0.483
Refusals $N = 410$	0.454	0.449

[a] Found by numerical integration for each unit.

RMSE value in the first row is for the 561 nonrespondents and in the second row for the 410 refusals.

For both definitions of nonresponse, the nonignorable model resulted in a smaller mean-squared error for imputed values than the ignorable model. This lends some support to the potential utility of using selection models with data like these. Of course, it is possible that using an ignorable model with more X variables, which are available, more extensive modeling of the X variables (more interactions, nonnormal residuals, or a different transformation of Y), would have resulted in even smaller RMSEs. In fact, David, Little, Samuhal, and Triest (1986) carefully study a more recent CPS exact match file and conclude that when using the full set of available X variables, there is no evidence of nonignorable nonresponse on income questions.

Estimates and Standard Errors for Average log(wage) for Nonrespondents in the Sample

GRZ considered multiple-imputation methods in addition to the single-imputation methods summarized in Table 6.3, which simply impute the posterior mean of Y_{mis} at $\theta = \hat{\theta}$. First, an ignorable multiple-imputation procedure was considered:

1. Draw 10 imputations from the posterior distribution of log(wage), where *a priori* $\gamma \equiv 0$, to create repeated multiple imputations for each nonrespondent.

Furthermore, two methods of nonignorable multiple imputation were considered:

2. Fix θ at its maximum-likelihood estimate $\hat{\theta}$, and draw 10 imputations from the conditional posterior distribution of log(wage) given $\theta = \hat{\theta}$, to create multiple imputations for each nonrespondent.
3. Draw 10 imputations from the posterior distribution of log(wage) to create repeated multiple imputations for each nonrespondent.

Table 6.4 displays repeated-imputation estimates and standard errors for the average log(wage) across nonrespondents in the sample for these five methods using both definitions of nonresponse. The two conditions without displayed values were not studied by GRZ. The single-imputation methods clearly overstate the accuracy of the imputed values since they lead to zero standard errors for the unknown sample average; this effect was also seen in the CPS example of Section 5.5.

TABLE 6.4. Repeated-imputation estimates (standard errors) for average log(wage) for nonrespondents in the sample under five imputation procedures.

	Ignorable Model		Nonignorable Selection Model		
	Fix θ at MLE Impute Mean[a]	Draw 10 Values from Posterior[b]	Fix θ at MLE Impute Mean[a]	Fix θ at MLE Draw 10 Values from Conditional Posterior[b]	Draw 10 Values from Posterior[b]
All nonrespondents ($N = 561$) True mean $Y = 9.558$	9.481 (0)	— —	9.512 (0)	9.519 (.018)	9.513 (.038)
Refusals ($N = 410$) True mean $Y = 9.571$	9.502 (0)	9.513 (.021)	9.571 (0)	— —	9.580 (.060)

[a] Posterior mean calculated for each unit by numerical integration.
[b] Reference distribution is t on 9 degrees of freedom.

The two repeated multiple-imputation procedures, which incorporate draws of θ from its posterior distribution, have the largest standard errors. When using the nonignorable model, the resulting inferences comfortably cover the true sample averages for both definitions of nonresponse. When using the properly implemented ignorable model, the estimate is nearly three $(9.571 - 9.513/.021)$ standard errors from the true sample average. The clear conclusion is that for these data, the ignorable model with normal residuals and two *a priori* zero regression coefficients can be improved by appending a nonignorable logistic response mechanism.

The method that fixes θ at its maximum-likelihood estimate before drawing 10 multiple imputations has a standard error, in the one case reported, nearly half as large as it should be, and as a consequence its estimate is more than two standard errors from the true sample average; the method should be rejected as a fully acceptable imputation procedure.

It is interesting to note that with the ignorable models, (a) fixing θ at $\hat{\theta}$ and imputing the mean, and (b) simulating the mean by 10 random draws, lead to similar estimates for the average Y for nonrespondents. An analogous conclusion holds for the estimates with the nonignorable methods. These results suggest that the posterior means of θ may be close to their maximum-likelihood estimates, and that 10 multiple imputations yield a fairly accurate estimate of the posterior mean of the average Y for nonrespondents.

Inferences for Population Mean log(wage)

GRZ do not calculate inferences for the population mean log(wage), but it is instructive to consider what additional information would be needed to do so. For example, assuming the standard interval would be used, we need in addition to the posterior mean and variance for the imputed average for respondents provided in Table 6.4: (1) the average log(wage) for the respondents in the sample so that we can calculate the posterior mean for the average log(wage) in the full sample and (2) the average variance of average log(wages) within the sample across the imputations.

6.6. EXTENSIONS TO SURVEYS WITH FOLLOW-UPS

A common practical device when faced with survey nonresponse that is feared to be nonignorable is to take a sample of the nonrespondents using an unconfounded probability sampling mechanism, and follow them up, in the sense that extra efforts are made to obtain responses from them. If the follow-up effort is entirely successful, all followed-up nonrespondents will respond and produce their values of Y_i. More commonly, there will be less than 100% response from the followed-up nonrespondents, the so-called hard-core nonrespondents still refusing to provide values of Y_i. In some surveys there will be waves of follow-up samples, each wave sampling some hard-core nonrespondents from the previous wave. Figure 6.1 depicts the process with two waves of follow-up sampling.

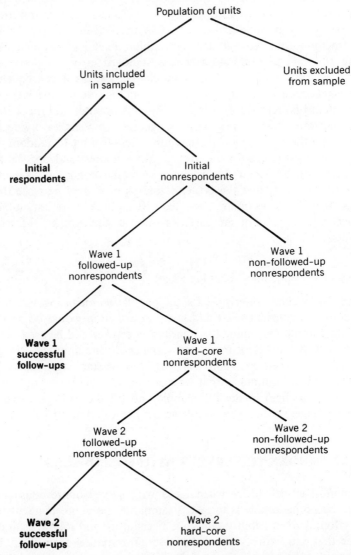

Figure 6.1. Schematic data structure with follow-up surveys of nonrespondents: boldface produces Y data.

EXTENSIONS TO SURVEYS WITH FOLLOW-UPS

The description just used of surveys with follow-ups is somewhat simplified in that with multivariate Y_i some followed-up nonrespondents may be respondents on some components of Y_i in the follow-up survey and hard-core nonrespondents on other components of Y_i. Nevertheless, to avoid cumbersome jargon and notation that conveys few new insights, we continue to use simplified descriptions, which are fully adequate only when Y_i is univariate.

Ignorable Nonresponse

If nonresponse is considered to be ignorable, then the methods of Chapter 5 can be immediately applied treating all values of Y observed in the original survey and in the follow-up survey as Y_{obs}, and multiply imputing the missing values, Y_{mis}, which are composed of the missing values from the non-followed-up nonrespondents and from the followed-up hard-core nonrespondents. The reason for expending effort and expense to collect follow-up data on nonrespondents, however, is usually that it is suspected that there are systematic differences between respondents and nonrespondents. Consequently, in cases with follow-up samples, it is rarely appropriate to assume ignorable nonresponse.

Nonignorable Nonresponse with 100% Follow-up Response

If nonresponse is nonignorable and the follow-up effort is entirely successful, the obviously appropriate group from which to draw imputed values is the group of followed-up nonrespondents, at least if this group is moderately large. That is, for the purpose of creating the multiple imputations for the non-followed-up nonrespondents, we can treat all initial nonrespondents as the entire sample, the followed-up nonrespondents as respondents, and the non-followed-up nonrespondents as nonrespondents, where the nonresponse is ignorable because the follow-up sampling mechanism is ignorable. With 100% follow-up response we can apply the methods of Chapter 5 to create multiple imputations for the non-followed-up nonrespondents using data from the followed-up nonrespondents.

The formal motivation for this type of imputation is easily described using the notation already developed in Section 6.2 for the case of univariate Y_i and no X_i. As there, let θ_1 be the parameter for the respondents' Y_i, and let θ_0 be the parameter for the nonrespondents' Y_i; the parameter of the distribution of the R_i is θ_R. When θ_0 is *a priori* independent of (θ_1, θ_R), it is easy to see that the posterior distribution of θ_0 is proportional to

$$\prod^{n_{01}} f_0(Y_i|\theta_0)\Pr(\theta_0) \qquad (6.6.1)$$

where the product is over the n_{01} followed-up nonrespondents. Repeated multiple imputations for the n_{00} missing values for the non-followed-up nonrespondents thus can be drawn in two steps:

1. Draw θ_0 from its posterior distribution, say θ_0^*.
2. Draw each of the n_{00} missing values of Y_i as i.i.d. from $f_0(Y_i|\theta_0 = \theta_0^*)$.

These two steps involve Y data only from followed-up nonrespondents because θ_0 is *a priori* independent of (θ_1, θ_R). When θ_0 is not *a priori* independent of (θ_1, θ_R), the posterior distribution of θ_0 involves Y data from initial respondents too; such models would be especially appropriate when n_{01} is small.

Example 6.3. 100% Follow-Up Response in a Simple Random Sample of Y_i

Suppose a random sample of n values of univariate Y_i is chosen, where there are n_1 respondents and n_0 nonrespondents. Furthermore, suppose a random sample of n_{01} of the n_0 nonrespondents is followed-up with complete success, where $n_{00} = n_0 - n_{01}$ is the number of non-followed-up nonrespondents. Multiple imputations for the n_{00} missing values are then created from the n_{01} followed-up nonrespondents using an ignorable repeated-imputation scheme that is proper for $\{\bar{y}, s^2(n^{-1} - N^{-1})\}$ as defined in Section 4.2. For instance, using the approximate Bayesian bootstrap introduced in Example 4.4, draw n_{01} values with replacement from the n_{01} followed-up nonrespondents' values, and then randomly draw the n_{00} missing values from the n_{01} values so chosen. Inferences for the population mean will be drawn using the standard complete-data statistics $\bar{y}$ and $s^2(n^{-1} - N^{-1})$, so that with large m the estimate for $\bar{y}$ is $E_*(\bar{y})$ with standard error $[E_*[s^2(n^{-1} - N^{-1})] + V_*(\bar{y})]^{1/2}$ where E_* and V_* refer to the mean and variance over the repeated imputations. Straightforward algebra shows that the expectation of $\bar{y}$ over repeated multiple imputations is

$$E_*(\bar{y}) = (n_1 \bar{y}_1 + n_0 \bar{y}_{01})/n \qquad (6.6.2)$$

where $\bar{y}_{01}$ is the average Y_i for the n_{01} followed-up nonrespondents; (6.6.2) is the standard randomization-based estimate of $\bar{y}$ with 100% follow-up response in a simple random sample [e.g., examine formula (13.14) from Cochran, 1977, p. 371]. Furthermore, when sample sizes are large,

$$V_*(\bar{y}) = s_{01}^2(n_0/n)(n_0/n_{01} - 1)/n \qquad (6.6.3)$$

and

$$E_*[s^2(n^{-1} - N^{-1})] = \left[(n_1 - 1)s_1^2 + (n_0 - 1)s_{01}^2 \right.$$
$$\left. + (n_1 n_0/n)(\bar{y}_1 - \bar{y}_{01})^2\right](n^{-1} - N^{-1}) \quad (6.6.4)$$

where s_{01}^2 is the sample variance of the n_{01} followed-up nonrespondents' values of Y_i. Examination of (6.6.3) shows that $V_*(\bar{y})$ estimates: (the population variance of Y_i among the nonrespondents) times (the fraction of the population units that are nonrespondents) times (the ratio of sample sizes for nonrespondents to followed-up nonrespondents minus one). Examination of (6.6.4) shows that $E_*[s^2(n^{-1} - N^{-1})]$ estimates (the population variance of the Y_i) times the usual standard error factor, $(n^{-1} - N^{-1})$. Thus, the repeated-imputation standard error for $\bar{Y}$, $[E_*[s^2(n^{-1} - N^{-1})] + V_*(\bar{y})]^{1/2}$, is essentially the same as that suggested by the double sampling approach [e.g., examine formula (13.15) from Cochran, 1977, p. 371].

Ignorable Hard-Core Nonresponse Among Follow-Ups

Suppose now that a random sample of nonrespondents is taken and followed up, but that this effort is less than 100% successful so that there are hard-core nonrespondents among the follow-ups. In some cases, it might be reasonable to assume because of the construction of the follow-up survey that this second stage of nonresponse is ignorable even though initial nonresponse is not ignorable. For example, perhaps no nonrespondents in the follow-up were refusals, and the reasons for follow-up nonresponse (e.g., not at home) are considered to be unrelated to the values of the outcome variable.

In such a case the observed values from the followed-up nonrespondents can be treated as if they were the result of 100% follow-up response, and so an ignorable method of Chapter 5 can be used, as was illustrated in Example 6.3. That is, we can simply act as if the hard-core nonrespondents were never sampled. Of course this would not be appropriate if hard-core nonresponse is thought to be nonignorable.

Nonignorable Hard-Core Nonresponse Among Follow-Ups

One way to create multiple imputations with hard-core nonresponse is to treat the follow-up survey of the nonrespondents as the entire survey. That is, treat the group of initial nonrespondents as the population, treat the group of followed-up nonrespondents as the sample, treat the successfully followed-up nonrespondents as the respondents, and treat the hard-core nonrespondents as the nonrespondents. Then apply the methods appropriate for nonignorable nonresponse discussed in the earlier sections of this chapter to the sample of followed-up nonrespondents to multiply impute missing values for the hard-core nonrespondents. Now handle the missing values from the non-followed-up nonrespondents as with 100% follow-up response. More explicitly, we take one draw of missing values for the hard-core nonrespondents using a nonignorable model, and then treating the drawn values as data just like the real data from successful followed-up nonrespondents, take one draw of the missing values for the non-followed-up nonrespondents using an ignorable model just as with a fully successful follow-up wave. These two steps create one draw of all missing values.

Waves of Follow-Ups

Waves of follow-ups are treated by analogy with the previous discussion. Start at the last follow-up wave and impute once for the hard-core nonrespondents, using, in general, a nonignorable model; then treating these imputed values as real, impute for the non-followed-up nonrespondents in this last wave using an ignorable model. Then move up one level of waves and impute for that wave's missing values. Continue until all values are imputed; repeat the procedure to create multiple imputations.

Each step can be carried out just as if it involved only one wave of follow-up sampling. The theoretical justification for this fact is closely connected with the results on monotone-distinct structure presented in Section 6.3 for the case without waves of nonresponse. The simplicity of the procedure requires the *a priori* independence of parameters for the different waves of nonresponse. In cases where the numbers of successfully followed up nonrespondents at some waves are small, it would be wise to try to borrow strength across the waves, and then the resulting procedure would be more complicated.

6.7. FOLLOW-UP RESPONSE IN A SURVEY OF DRINKING BEHAVIOR AMONG MEN OF RETIREMENT AGE

We illustrate the use of multiple imputation when there exist followed-up nonrespondents using data from a mailed survey of alcohol consumption among men of retirement age. This example is from Glynn, Laird, and Rubin (1986) and involves men participating in an ongoing study, the Normative Aging Study, who were between 50 and 69 years old in 1982 with known retirement status.

A SURVEY OF DRINKING BEHAVIOR

The Data Base

A total of 1384 men in the Normative Aging Study with known retirement status were sent the mailed questionnaire; 1272 provided information on drinking behavior [$Y = \log(1 + \text{number of drinks per day})$], whereas 112 were initial nonrespondents. Of these 112 initial nonrespondents, 38 eventually provided values of Y. Table 6.5 provides some basic statistics for this data set. We will consider the follow-up survey to be entirely successful in the sense that we will draw multiple imputations for the $74 (= 112 - 38)$ missing Y values using the followed-up nonrespondents data and an ignorable procedure.

The Modeling Task

The model for the initial nonrespondents is a normal linear regression model for Y on X = (constant, retirement status, birth year, retirement status × birth year), where the standard noninformative prior was used for the regression coefficients, β, and the conditional variance σ^2. Thus, this model is formally identical to the ignorable model of Example 5.1, where the 112 initial nonrespondents are treated as the sample and the 38 followed-up nonrespondents are treated as the respondents.

The Estimation Task

The estimation task can be summarized using the usual least-squares statistics for the regression of Y on X for the 38 followed-up nonrespondents. For purposes of comparison, Table 6.6 also provides the least-squares summaries for the corresponding regression of Y on X for the 1272 initial respondents. The posterior distribution for the nonrespondents' variance, σ^2, is $34(.565)^2/\chi^2_{34}$, and given σ^2 the posterior distribution for the nonrespondents' regression coefficient, β, is normal with mean given by the least-squares estimate and variance–covariance matrix given by $\sigma^2(X'X)^{-1}$. This completes the estimation task.

The Imputation Task

The imputation task is very straightforward and is repeated five times to create five imputations:

1. Draw (β, σ^2) from its posterior distribution formulated in the estimation task; say β_*, σ^2_*.

TABLE 6.5. Mean alcohol consumption level and retirement status for respondents and nonrespondents within birth cohort: Data from 1982 Normative Aging Study drinking questionnaire.[a]

	Respondents					Nonrespondents				
		Percent Retired	log(1 + drinks/day)				Percent Retired		log(1 + drinks/day)	
			Mean	S.D.				Number of Follow-ups	Follow-ups	
Birth Year	N					N			Mean	S.D.
1928–1932	329	7.6	0.79	0.57		61	8.2	17	0.52	0.42
1922–1927	398	21.6	0.69	0.57		21	9.5	5	0.56	0.61
1918–1922	334	43.4	0.71	0.58		20	25.0	10	0.84	0.64
1913–1917	211	79.1	0.64	0.54		10	100.0	6	0.86	0.82
Total	1272					112		38		

[a] Source: Glynn, Laird, and Rubin (1986), Table 6.

2. Draw the 74 non-followed-up nonrespondents' values of Y_i independently as

$$Y_i = X_i \beta_* + \sigma_* z_i, \quad i \in mis$$

where the z_i are i.i.d. standard normal deviates.

Table 6.7 provides the five values of (β, σ^2) that were drawn and Table 6.8 provides the five values of Y_{mis}.

TABLE 6.6. Summary of least-squares estimates of the regression of log(1 + drinks / day) on retirement status (0 = working, 1 = retired), birth year, and retirement status × birth year interaction.[a]

Variable	Estimates Based on 1272 Respondents to the 1982 Questionnaire		Estimates Based on 38 Nonrespondents from Whom log(1 + drinks/day) was Later Obtained	
	Coefficient	Standard Error	Coefficient	Standard Error
Retirement status	0.334	0.166	−0.589	0.900
Birth year	0.0144	0.00434	−0.0292	0.0252
Retirement × birth year	−0.0141	0.00747	0.0372	0.0385
Constant	0.00161	0.0159	−0.0280	0.0948
	$R^2 = 0.0091$ Standard Deviation = 0.568		$R^2 = 0.122$ Standard Deviation = 0.565	

[a]*Source*: Glynn, Laird and Rubin (1986), Table 7.

TABLE 6.7. Five values of regression parameters for nonrespondents drawn from their posterior distribution.[a]

	Imputation Number				
	1	2	3	4	5
Retirement status	−0.055	0.38	−0.10	−1.95	−1.03
Birth year	0.0036	−0.0034	0.0067	−0.088	−0.017
Retirement × birth year	−0.026	0.016	0.029	0.087	0.063
Constant	−0.010	−0.13	−0.037	−0.104	−0.021
Standard deviation	0.51	0.55	0.59	0.58	0.60

[a]*Source*: Glynn, Laird, and Rubin (1986), Table 18.

TABLE 6.8. Five imputed values of log(1 + drinks / day) for each of the 74 non-followed-up nonrespondents.[a]

	Imputation Number						Imputation Number				
	1	2	3	4	5		1	2	3	4	5
1	0.0	0.0	1.6	0.8	0.6	38	1.7	0.7	0.9	0.0	0.1
2	0.0	0.9	1.0	0.0	0.7	39	1.3	0.0	0.7	0.1	0.0
3	0.0	0.9	1.1	0.0	0.9	40	0.6	0.1	1.0	0.7	0.9
4	1.4	0.3	1.3	0.0	0.4	41	0.4	1.2	0.0	0.0	0.0
5	1.0	0.5	0.6	0.6	1.7	42	0.2	0.0	0.6	0.1	1.0
6	0.3	0.9	0.7	1.7	0.0	43	0.3	0.8	0.6	0.0	1.1
7	1.5	1.0	0.6	1.8	2.0	44	1.0	0.0	1.0	0.0	0.4
8	0.8	1.2	0.6	0.6	0.7	45	0.7	0.0	0.7	0.4	0.4
9	0.0	0.0	0.6	1.5	0.9	46	0.4	0.0	1.5	0.2	0.6
10	0.0	0.4	0.0	0.5	1.4	47	1.6	0.4	0.1	0.5	0.6
11	0.3	0.3	0.0	0.5	1.5	48	0.7	1.0	1.2	1.0	0.6
12	1.0	0.5	0.8	1.7	0.1	49	0.9	0.3	1.0	0.0	1.8
13	0.0	0.5	1.5	0.0	1.3	50	0.2	0.0	0.1	0.0	0.6
14	1.4	0.1	0.0	0.0	0.6	51	0.8	1.3	0.7	0.2	0.2
15	0.6	1.6	0.0	2.1	0.9	52	0.0	0.5	0.9	0.4	0.3
16	0.6	0.2	0.0	0.0	0.3	53	0.0	0.9	1.1	1.5	0.5
17	1.4	1.5	0.3	0.5	1.7	54	0.0	0.0	0.8	0.7	0.9
18	1.1	0.5	1.7	1.1	0.5	55	0.5	0.0	0.7	0.1	0.9
19	1.1	0.0	0.7	1.3	0.4	56	0.0	0.2	0.1	0.0	1.5
20	0.3	0.0	1.1	0.0	0.7	57	0.0	0.0	0.2	0.0	0.0
21	1.7	0.0	0.5	0.4	0.4	58	0.9	0.8	0.3	1.2	0.7
22	0.3	0.0	0.0	0.0	1.1	59	1.0	0.5	1.5	0.6	1.2
23	0.3	0.7	0.0	0.0	0.2	60	0.4	1.4	0.2	0.0	0.0
24	0.0	0.0	0.3	0.6	0.8	61	0.6	0.4	0.4	0.4	0.2
25	0.4	0.0	0.7	2.4	0.9	62	1.0	0.3	0.7	0.8	0.3
26	1.0	0.5	1.0	0.0	0.4	63	0.0	0.3	0.1	0.4	0.4
27	0.5	0.4	0.8	0.0	0.5	64	1.2	0.2	0.0	0.0	0.0
28	1.0	0.3	0.5	0.2	0.1	65	0.0	0.7	1.1	0.4	0.0
29	0.8	0.9	0.7	0.1	2.1	66	0.6	0.0	0.0	0.0	0.0
30	0.2	0.0	0.0	2.0	1.4	67	1.1	1.2	1.0	0.4	0.0
31	0.4	0.8	0.0	0.3	0.2	68	0.4	0.4	1.0	0.9	0.5
32	0.8	0.1	0.1	0.0	1.3	69	1.6	0.2	1.3	1.6	0.6
33	1.0	0.5	0.0	0.7	0.3	70	0.5	2.0	1.0	0.5	1.3
34	1.1	1.2	0.0	1.6	1.0	71	0.8	0.9	0.1	0.0	0.0
35	0.8	0.6	0.5	0.5	0.9	72	0.8	1.2	0.6	0.5	0.7
36	0.3	0.0	1.1	0.9	1.1	73	0.6	0.6	0.4	0.0	1.0
37	0.8	0.0	1.8	0.0	0.6	74	0.0	1.1	0.0	1.6	1.1

[a]*Source*: Glynn, Laird, and Rubin (1986), Table 19.

TABLE 6.9. Sets of least-squares estimates from the five data sets completed by imputation.[a]

	Imputation Number				
	1	2	3	4	5
Retirement status	0.243	0.237	0.244	0.126	0.217
Birth year	0.0106	0.00953	0.01180	0.00594	0.01025
Retirement × birth year	−0.00907	−0.00854	−0.00917	−0.00420	−0.00790
Constant	−0.00620	−0.0133	−0.00658	−0.01112	−0.00321
Standard deviation	0.565	0.567	0.565	0.574	0.567

[a]*Source*: Glynn, Laird, and Rubin (1986), Table 20.

Inference for the Effect of Retirement Status on Drinking Behavior

In the absence of nonresponse, inferences about the effect of retirement status on drinking behavior adjusting for age (birth year) would have been drawn by the investigators involved by regressing Y_i on X_i using ordinary least squares. With multiple imputation, the same analysis is now carried out on each of the five data sets completed by imputation; the five resulting estimates of (β, σ^2) are given in Table 6.9. The usual repeated-data estimates and standard errors are formed from the five complete-data estimates and standard errors; the results are summarized in Table 6.10.

The results suggest a possible relationship of retirement status and drinking behavior after adjusting for age. It is interesting to note that the fractions of missing information, as displayed in Table 6.10, can be larger

TABLE 6.10. Repeated-imputation estimates, standard errors, and percentages of missing information for the regression of log(1 + drinks / day) on retirement status, birth year, and retirement status × birth year interaction.[a]

	Estimate	Standard Error	Percent of Missing Information
Retirement status	0.213	0.168	9.1
Birth year	0.00962	0.00480	22.5
Retirement × birth year	−0.00778	0.00747	7.7
Constant	−0.00809	0.0159	6.7
Standard deviation	0.568		

[a]*Source*: Glynn, Laird, and Rubin (1986), Table 21.

than the fraction of Y_i values that are missing (74/1384 = 5%). This occurs because the nonrespondents' and respondents' regressions of Y_i on X_i and marginal distributions of X_i are different (see Table 6.6 for the regressions of Y_i on X_i, and Table 6.5 for the marginal distributions of X_i).

PROBLEMS

1. Suppose that (Y_i, X_i) have a multivariate normal distribution with common correlation among all pairs of variables. Describe the modeling, estimation, and imputation tasks with nonignorable nonresponse.
2. Describe a log-linear formulation analogous to the one in Problem 1 when (Y_i, X_i) are categorical.
3. Suppose that there are K patterns of missing values across the units included in the survey, and let $\theta^{(k)}$, $k = 1, \ldots, K$ be the parameter governing the distribution of Y_i given X_i for each pattern, where the Kth pattern has no missing data. Consider models that posit

$$\Pr(\theta^{(1)}, \ldots, \theta^{(K-1)} | \theta^{(K)}) = \prod_{k=1}^{K-1} \Pr(\theta^{(k)} | \theta^{(K)}).$$

Discuss the possible use of such models in practice (hint: when $K = 2$, consider the model in Section 6.4 and extensions).

4. Discuss possible pathways for building nonignorable models that incorporate realistic prior information other than the models discussed in this chapter (e.g., see Rubin, 1978a). In particular, consider the role of exchangeability assumptions across variables and across patterns of nonresponse.
5. In the context of the selection model of Example 6.2, investigate the normality of the posterior distribution of θ. Also comment on the use of standard results (e.g., at $\rho = 0$, μ and σ are *a posteriori* independent; what happens when $\rho \neq 0$?).
6. In the context of the simple case of Section 6.2, suppose the n_0 values to impute are drawn with replacement with probability proportional to the value to be imputed from n_1 values drawn using the approximate Bayesian bootstrap. Under what explicit nonignorable model is this implicit procedure approximately appropriate? Evaluate the performance of the repeated-imputation inference for $\overline{Y}$ based on the standard complete-data statistics.

PROBLEMS

7. Carry out a frequentist evaluation of the mixture model and selection model approaches of Example 6.1 and 6.2 in specific cases.
8. Prove Result 6.1.
9. Prove Result 6.2.
10. Describe a realistic example in which it would be unwise to assume $\theta_{Y|XR}$ is *a priori* independent of $(\theta_X, \theta_{R|X})$, and specify a more reasonable prior distribution in this example.
11. Describe the estimation and imputation tasks for the model proposed in Problem 10.
12. Prove Result 6.3.
13. Prove Result 6.4.
14. Prove Result 6.5.
15. Develop precise notation for the case of monotone missingness with nonignorable nonresponse.
16. Prove Result 6.6.
17. Suppose $\Pr(X, Y, R)$ is a selection model. Further suppose that given $(X_i, Y_{i,inc}, \theta_{R|XY})$, the probability of nonresponse on components of Y_i included in the sample does not depend on components of Y_i excluded from the sample:

$$\Pr(R_{i,inc}|X_i, Y_i, \theta_{R|XY}) = \Pr(R_{i,inc}|X_i, Y_{i,inc}, \theta_{R|XY}).$$

(a) Comment on this assumption. Interpret it in the context of a specific example, and consider how it relates to Examples 3.6 and 3.7.

(b) Write the posterior distribution of Y_{exc} as

$$\Pr(Y_{exc}|X, Y_{inc}, R_{inc}) = \frac{\int\int \Pr(X, Y, R|\theta)\Pr(\theta)\, d\theta\, dR_{exc}}{\int\int\int \Pr(X, Y, R|\theta)\Pr(\theta)\, d\theta\, dR_{exc}\, dY_{exc}},$$

and integrate over R_{exc} and $\theta = (\theta_{XY}, \theta_{R|XY})$ to obtain

$$\Pr(Y_{exc}|X, Y_{inc}, R_{inc}) = \frac{\int \prod_{i=1}^{N} f_{XY}(X_i, Y_i|\theta_{XY})\Pr(\theta_{XY})\, d\theta_{XY}}{\int\int \prod_{i=1}^{N} f_{XY}(X_i, Y_i|\theta_{XY})\Pr(\theta_{XY})\, d\theta_{XY}\, dY_{exc}}.$$

(c) Conclude that under the assumptions of the problem the complete-data and completed-data posterior distributions of any Q are equal.

18. Prove the claim following Result 6.6 concerning selection modeling and monotone missingness. Also, are the conditions stated there *necessary* for the estimation and imputation tasks to reduce to a sequence of univariate estimation and imputation tasks when using the selection modeling approach?

19. Accepting the model used in Section 6.4, justify the estimation and imputation tasks described there. Propose an alternative model (e.g., one based on a logistic regression model), and comment on the relative simplicity of the estimation and imputation tacks.

20. Criticize the model used in Section 6.5. How valid is step 1 in the imputation task, and should the results be trusted (hint: see Problem 5)? Propose an alternative model and discuss relative strengths and weaknesses.

21. Describe how to use all data in the Example of Section 6.5, CPS and administrative, to create multiple imputations for CPS missing values. Discuss whether such a procedure could be used in practice.

22. Develop precise notation extending the notation already established to handle waves of follow-up sampling and hard-core nonresponse with multivariate Y_i.

23. Prove that the posterior distribution of θ_0 is proportional to (6.6.1) in the simple case of that equation.

24. For the case of Example 6.3, describe the steps of a multiple-imputation scheme assuming nonresponse is ignorable.

25. Prove the large-sample results of Example 6.3 (hint: let $\bar{y}_{*I} = (n_1/n)\bar{y}_1 + (n_0/n)[(n_{01}/n)\bar{y}_{01} + (n_{00}/n)\bar{y}_{00I}]$, and $s_{I*}^2 = [(n_1 - 1)s_1^2 + (n_0 - 1)s_{0I}^2 + n_1 n_0 (\bar{y}_1 - \bar{y}_0)^2]/(n-1)$, $s_{0I}^2 = \cdots$)

26. Derive small-sample expressions for $E_*(\bar{y})$, $E_*(s^2)$, and $V_*(\bar{y})$ in the context of Example 6.3 for

(a) the Bayesian bootstrap,

(b) the approximate Bayesian bootstrap,

(c) fully normal imputation,

(d) the mean and variance adjusted hot-deck.

27. Propose a method of multiple imputation for the case of Example 6.3 when n_{01} is small, and derive its frequentist properties.

28. Outline methods for multiple imputation in the case of Example 6.3 extended to include hard-core nonresponse.

29. Provide a flow chart for a general multiple-imputation scheme when there are waves of follow-up sampling and parameters for the waves are *a priori* independent. Prove that the procedure can be applied independently across waves.

PROBLEMS

30. Comment on the relative ease of deriving inferences analytically with waves of followed-up nonrespondents (e.g., extending results in Cochran, 1977) versus relying on the general theory of multiple imputation to simulate results.

31. Review the statistical literature on methods for nonignorable nonresponse (hint: check Little, 1982, 1983b; Morris, 1983b and the bibliographies in the NAS volumes on incomplete data).

32. Discuss general ways of borrowing strength across waves of nonrespondents when creating multiple imputations.

33. *Undercoverage.* Suppose we have a sample of n values of Y_i, but N is not known. Describe how to create multiple imputations for the unobserved $N - n$ values of Y_i when the Y_i $i = 1, \ldots, N$ are i.i.d. $N(0, 1)$ and observed only if $Y_i > c$, c known. Extend these results to Y_i i.i.d. $N(\mu, \sigma^2)$ where μ, σ^2 are to be estimated and Y_i is observed only if $> c_1$ or $< c_2$.

34. *Undercoverage continued.* Extend the results of Problem 32 to cases where c_1 and c_2 must be estimated. Consider further extensions to cases where Y_i is always observed if $c_1 < Y_i < c_2$ and possibly observed otherwise. Compare results with methods suggested in Wachter and Trussel (1982) for estimating μ, σ^2.

35. *Undercoverage continued.* Comment on the sensitivity of results to the normality assumption and the general applicability of such methods for handling undercoverage in practice. Suggest better methods for handling undercoverage.

References

Afifi, A. A., and Elashoff, R. M. (1969). Missing observations in multivariate statistics III: Large sample analysis of simple linear regression. *Journal of the American Statistical Association* **64**, 337–358.

Althauser, R., and Rubin, D. B. (1971). The computerized construction of a matched sample. *American Journal of Sociology* **76**, 325–346.

Aziz, F., Kilss, B., and Scheuren, F. (1978). 1973 Current Population Survey—administrative record exact match file codebook. Report No. 4, Studies from Interagency Data Linkages. Washington, D.C.: Social Security Administration.

Beale, E. M. L., and Little, R. J. A. (1975). Missing values in multivariate analysis. *Journal of the Royal Statistical Society Series* **B37**, 129–145.

Beaton, A. E. (1964). The use of special matrix operations in statistical calculus. (ETS RB-64-51). Princeton, N.J.: Educational Testing Service.

Berndt, E. K., Hall, B. H., Hall, R. E., and Hausman, J. A. (1974). Estimation and inference in nonlinear structure models. *Annals of Economic and Social Measurement* **4**, 653–665.

Binder, D. A. (1982). Nonparametric Bayesian models for samples from finite populations. *Journal of the Royal Statistical Society* **B44**, 388–393.

Box, G. E. P., and Tiao, G. C. (1973). *Bayesian Inference in Statistical Analysis.* Reading, Mass: Addison-Wesley.

Bradburn, N. M., and Sudman, S. (1979). *Improving Interview Method and Questionnaire Design.* San Francisco: Jossey-Bass.

Cochran, W. G. (1968). The effectiveness of adjustment by subclassification in removing bias in observational studies. *Biometrics* **24**, 295–313.

Cochran, W. G. (1977). *Sampling Techniques.* New York: Wiley.

Cochran, W. G. (1978). Laplace's ratio estimator. In *Contributions to Survey Sampling and Applied Statistics.* H. A. David (ed.). New York: Academic Press, pp. 3–10.

Cochran, W. G., and Rubin, D. B. (1973). Controlling bias in observational studies: a review. *Sankhya* **A35**, 417–446.

Cox, D. R. (1958). *Planning of Experiments.* New York: Wiley.

Cox, D. R., and Hinkley, D. V. (1974). *Theoretical Statistics.* London: Chapman and Hall.

David, M., Little, R. J. A., Samuhal, M. E., and Triest, R. K. (1986). Alternative methods for CPS income imputation. *Journal of the American Statistical Association* **81**, 29–41.

REFERENCES

Dawid, A. P. (1979). Conditional independence in statistical theory. *Journal of the Royal Statistical Society* **B41**, 1–31.

Dawid, A. P. (1982). The well-calibrated Bayesian. *Journal of the American Statistical Association* **77**, 605–613.

Dempster, A. P. (1969). *Elements of Continuous Multivariate Analysis*. Reading, Mass.: Addison-Wesley.

Dempster, A. P., Laird, N., and Rubin, D. B. (1977). Maximum likelihood from incomplete data via the EM algorithm. *Journal of the Royal Statistical Society* **B39**, 1–38.

Dempster, A. P., Laird, N., and Rubin, D. B. (1980). Interatively reweighted least squares for linear regression when errors are normal/independent distributed. *Multivariate Analysis* **V**, 35–57.

Dempster, A. P., Rubin, D. B., and Tsutakawa, R. K. (1981). Estimation in covariance components models. *Journal of the American Statistical Association* **76**, 341–353.

Diaconis, P. (1977). Finite forms of deFinetti's theorem on exchangeability. *Synthese* **36**, 271–281.

Dijkstra, W., and Vander Zouwen, J. (1982) (eds.). *Response Behaviors in the Survey Interview*. New York: Academic Press.

Dillman, D. A. (1978). *Mail and Telephone Surveys. The Total Design Method*. New York: Wiley.

Efron, B. (1975). The efficiency of logistic regression compared to normal discriminant analysis. *Journal of the American Statistical Association* **70**, 892–898.

Efron, B., and Morris, C. N. (1975). Data analysis using Stein's estimator and its generalizations. *Journal of the American Statistical Association* **70**, 311–319.

Ericson, W. A. (1969). Subjective Bayesian models in sampling finite populations. *Journal of the Royal Statistical Society* **B31**, 195–233.

Feller, W. (1966). *An Introduction to Probability Theory and Its Applications, Volume II*. New York: Wiley.

Fisher, R. A. (1935). *The Design of Experiments*, London: Oliver and Boyd.

Ford, B. L. (1983). An overview of hot-deck procedures. In *Incomplete Data in Sample Surveys, Volume 2, Theory and Bibliographies*. W. G. Madow, I. Olkin, and D. B. Rubin, (eds.). New York: Academic Press, pp. 185–207.

Glynn, R., Laird, N., and Rubin, D. B. (1986). Selection modelling versus mixture modelling with nonignorable nonresponse. In *Drawing Inferences from Self-Selected Samples*. H. Wainer (ed.). New York: Springer-Verlag, pp. 115–142.

Greenlees, J. S., Reece, W. S., and Zieschang, K. D. (1982a). Imputation of missing values when the probability of response depends on the variable being imputed. *Journal of the American Statistical Association* **77**, 251–261.

Greenlees, J. S., Reece, W. S., and Zieschang, K. D. (1982b). Imputing missing values when response behavior is nonignorable. Presented at the 1982 annual meeting of the American Statistical Association.

Groves, G. M., and Kahn, R. L. (1979). *Survey by Telephone—A National Comparison with Personal Interviews*. New York: Academic Press.

Hajek, J. (1960). Limiting distributions in simple random sampling from a finite population. *Pub. Math. Inst. Hungarian Acad. Sci.* **5**, 361–374.

Hansen, M. H., Hurwitz, W. N., and Madow, W. G. (1953). *Sample Survey Methods and Theory, Volumes 1 and 2*. New York: Wiley.

Hansen, M. H., Madow, W. G., and Tepping, J. (1983). An evaluation of model-dependent and probability-sampling inferences in sample surveys. *Journal of the American Statistical Association* **78**, 776–807.

Hartley, H. O. (1958). Maximum likelihood estimation from incomplete data. *Biometrics* **27**, 783–823.

Heckman, J. (1976). The common structure of statistical models of truncation, sample selection and limited dependent variables and a simple estimator for such models. *Annals of Economic and Social Measurement* **5**, 475–492.

Heitjan, D. F., and Rubin, D. B. (1986). Inference from coarse data using multiple imputation. *Proceedings of the 18th Symposium on the Interface of Computer Science and Statistics*. T. Boardman (ed.). Washington, D.C.: American Statistical Association, pp. 138–143.

Herzog, T. N. (1980). Multiple imputation modeling for individual social security benefit amounts, Part II. *Proceedings of the Survey Research Methods Section of the American Statistical Association*, 404–407.

Herzog, T. N., and Lancaster, C. (1980). Multiple imputation modeling for individual social security benefit amounts, Part I. *Proceedings of the Survey Research Methods Section of the American Statistical Association*, 398–403.

Herzog, T. N., and Rubin, D. B. (1983). Using multiple imputations to handle nonresponse in sample surveys. In *Incomplete Data in Sample Surveys, Volume 2, Theory and Bibliographies*. W. G. Madow, I. Olkin, and D. B. Rubin (eds.). New York: Academic Press, pp. 209–245.

Hewitt, E., and Savage, L. J. (1956). Symmetric measures on cartesian products. *Transactions of the American Mathematical Society* **80**, 470–501.

Holt, D., and Smith, T. W. F. (1979). Post stratification. *Journal of the Royal Statistical Society* **A142**, 33–46.

Huber, P. J. (1967). The behavior of maximum likelihood estimates under nonstandard conditions. *Proceedings of the Fifth Berkeley Symposium on Mathematical Statistics and Probability*. Berkeley: University of California Press, 221–233.

James, W., and Stein, C. (1961). Estimation with quadratic loss. *Proceedings of the Fourth Berkeley Symposium on Mathematical Statistics and Probability*. Berkeley: University of California Press, 361–379.

Kish, L. (1965). *Survey Sampling*. New York: Wiley.

Lancaster, C. (1979). Determining social security recipiency for imputation of nonresponse in the CPS. Social Security Administration Memorandum SRV-53. Washington, D.C., January 26.

Li, K. H. (1985). Hypothesis Testing in Multiple Imputation—with Emphasis on Mixed-up Frequencies in Contingency Tables. Ph.D. Thesis, Department of Statistics, University of Chicago.

Lillard, L., Smith, J. P., and Welch, F. (1986). What do we really know about wages: the importance of nonreporting and census imputation. *Journal of Political Economy* **94**, 489–506.

Lindley, D. V. (1972). *Bayesian Statistics. A Review*. Philadelphia: Society for Industrial and Applied Mathematics.

Lindley, D. V., and Smith, A. F. M. (1972). Bayes estimates for the linear model. *Journal of the Royal Statistical Society*, **B34**, 1–41.

Little, R. J. A. (1982). Models for nonresponse in sample surveys. *Journal of the American Statistical Association* **77**, 237–250.

REFERENCES

Little, R. J. A. (1983a). Superpopulation models for nonresponse—the ignorable case. In *Incomplete Data in Sample Surveys, Volume 2, Theory and Bibliographies.* W. G. Madow, I. Olkin, and D. B. Rubin (eds.). New York: Academic Press, pp. 341–382.

Little, R. J. A. (1983b). Superpopulation models for nonresponse—the nonignorable case. In *Incomplete Data in Sample Surveys, Volume 2, Theory and Bibliographies.* W. G. Madow, I. Olkin, and D. B. Rubin (eds.). New York: Academic Press, pp. 383–413.

Little, R. J. A. (1983c). Discussion of paper by Hasselblad, Creason and Stead. In *Incomplete Data in Sample Surveys, Volume 3, Proceedings of the Symposium.* W. G. Madow, and I. Olkin (eds.). New York: Academic Press, pp. 206–208.

Little, R. J. A., and Rubin, D. B. (1987). *Statistical Analysis with Missing Data.* New York: Wiley.

Little, R. J. A., and Schluchter, M. D. (1985). Maximum likelihood estimation for mixed continuous and categorical data with missing values. *Biometrika* **72**, 497–512.

Madow, W. G. (1948). On the limiting distributions of estimates based on samples from finite universes. *Annals of Mathematical Statistics* **19**, 535–545.

Madow, W. G., Nisselson, H., and Olkin, I. (1983). *Incomplete Data in Sample Surveys, Volume 1, Report and Case Studies.* New York: Academic Press.

Madow, W. G., and Olkin, I. (1983). *Incomplete Data in Sample Surveys. Volume 3, Proceedings of the Symposium.* New York: Academic Press.

Madow, W. G., Olkin, I., and Rubin, D. B. (1983). *Incomplete Data in Sample Surveys, Volume 2, Theory and Bibliographies.* New York: Academic Press.

Marini, M. M., Olsen, A. R., and Rubin, D. B. (1980). Maximum likelihood estimation in panel studies with missing data. *Sociological Methodology*, 314–357.

Mitra, S. K., and Pathak, D. K. (1984). The nature of simple random sampling. *Annals of Statistics* **12**, 1536–1542.

Morris, C. N. (1983a). Parametric empirical Bayes inference: theory and applications. *Journal of the American Statistical Association* **78**, 47–55.

Morris, C. N. (1983b). Nonresponse issues in public policy experiments, with emphasis on the health insurance study. In *Incomplete Data in Sample Surveys, Volume 3, Proceedings of the Symposium.* W. G. Madow, and I. Olkin (eds.). New York: Academic Press, pp. 313–325.

Nelson, F. D. (1977). Censored regression models with unobserved, stochastic censoring thresholds. *Journal of Econometrics* **6**, 309–327.

Neyman, J. (1934). On the two different aspects of the representative method: The method of stratified sampling and the method of purposive selection. *Journal of the Royal Statistical Society* **A97**, 558–606.

Oh, H. L., and Scheuren, F. J. (1980). Estimating the variance impact of missing CPS income data. *Proceedings of the Survey Research Methods Section of the American Statistical Association*, 408–415.

Oh, H. L., and Scheuren, F. J. (1983). Weighting adjustments for unit non-response. In *Incomplete Data in Sample Surveys, Volume 2, Theory and Bibliographies.* W. G. Madow, I. Olkin, and D. B. Rubin (eds.). New York: Academic Press, pp. 143–184.

Okner, B. (1974). Data matching and merging: an overview. *Annals of Economic and Social Measurement* **3**, 347–352.

Orchard, T., and Woodbury, M. A. (1972). A missing information principle: theory and applications. *Proceedings of the Sixth Berkeley Symposium on Mathematical Statistics and Probability.* Berkeley: University of California Press, 697–715.

Platek, R., and Grey, G. B. (1983). Imputation methodology: total survey error. In *Incomplete Data in Sample Surveys, Volume 2, Theory and Bibliographies*. W. G. Madow, I. Olkin, and D. B. Rubin, (eds.). New York: Academic Press, pp. 249-333.

Pratt, J. W. (1965). Bayesian interpretation of standard inference statements. *Journal of the Royal Statistical Society* **B27**, 169-203.

Raghunathan, T. E. (1987). Large sample significance levels from multiply-imputed data. Ph.D. Thesis, Department of Statistics, Harvard University.

Rodgers, W. L. (1984). An evaluation of statistical matching. *Journal of Business and Economic Statistics* **2**, 91-102.

Rosenbaum, P. R., and Rubin, D. B. (1983). The central role of the propensity score in observational studies for causal effects. *Biometrika* **70**, 41-55.

Rosenbaum, P. R., and Rubin, D. B. (1985). Constructing a control group using multivariate matched sampling incorporating the propensity score. *The American Statistician* **39**, 33-38.

Rosenthal, R., and Rosnow, R. L. (1975). *The Volunteer Subject*. New York: Wiley.

Rubin, D. B. (1973). Matching to remove bias in observational studies. *Biometrics* **29**, 159-183. Printer's correction note **30**, 728.

Rubin, D. B. (1974). Characterizing the estimation of parameters in incomplete data problems. *Journal of the American Statistical Association* **69**, 467-474.

Rubin, D. B. (1976a). Inference and missing data. *Biometrika* **63**, 581-592.

Rubin, D. B. (1976b). Multivariate matching methods that are equal percent bias reducing, I: some examples. *Biometrics* **32**, 109-120. Printer's correction note 955.

Rubin, D. B. (1977a). Formalizing subjective notions about the effect of nonrespondents in sample surveys. *Journal of the American Statistical Association* **72**, 538-543.

Rubin, D. B. (1977b). The design of a general and flexible system for handling non-response in sample surveys. Manuscript prepared for the U.S. Social Security Administration, July 1, 1977.

Rubin, D. B. (1978a). Multiple imputations in sample surveys—a phenomenological Bayesian approach to nonrepsonse. *Proceedings of the Survey Research Methods Section of the American Statistical Association*, 20-34. Also in *Imputation and Editing of Faulty or Missing Survey Data*. U.S. Dept. of Commerce, Bureau of the Census, 1-23.

Rubin, D. B. (1978b). The phenomenological Bayesian perspective in sample surveys from finite populations: foundations. *Imputation and Editing of Faulty or Missing Survey Data*. U.S. Dept. of Commerce, Bureau of the Census, 10-18.

Rubin, D. B. (1979a). Illustrating the use of multiple imputations to handle nonresponse in sample surveys. *Proceedings of the 42nd Session of the International Statistical Institute, Book 2*, 517-532.

Rubin, D. B. (1979b). Using multivariate matched sampling and regression adjustment to control bias in observational studies. *Journal of the American Statistical Association* **74**, 318-328.

Rubin, D. B. (1980a). *Handling Nonresponse in Sample Surveys by Multiple Imputations*. U.S. Dept. of Commerce, Bureau of the Census Monograph.

Rubin, D. B. (1980b). Discussion of "Randomization analysis of experimental data in the Fisher randomization test" by Basu. *Journal of the American Statistical Association* **75**, 591-593.

Rubin, D. B. (1980c). Bias reduction using Mahalanobis' metric matching. *Biometrics* **36**, 295-298. Printer's correction p. 296 (5, 10) = 75%.

REFERENCES

Rubin, D. B. (1980d). Introduction to papers on the CPS hot deck: Evaluations and alternatives and related research. *Economic and Demographic Statistics. U.S. Department of Health and Human Services*, Social Security Administration, Office of Research and Statistics, 67–70.

Rubin, D. B. (1981a). The Bayesian bootstrap. *The Annals of Statistics* **9**, 130–134.

Rubin, D. B. (1981b). Estimation in parallel randomized experiments. *Journal of Educational Statistics* **6**, 377–400.

Rubin, D. B. (1983a). Conceptual issues in the presence of nonresponse. In *Incomplete Data in Sample Surveys, Volume 2, Theories and Bibliographies*. W. G. Madow, I. Olkin, and D. B. Rubin (eds.). New York: Academic Press, pp. 123–142.

Rubin, D. B. (1983b). A case-study of the robustness of Bayesian/likelihood methods of inference: estimating the total in a finite population using transformations to normality. In *Scientific Inference, Data Analysis and Robustness*. G. E. P. Box, T. Leonard and C. F. Wu (eds.). New York: Academic Press, pp. 213–244.

Rubin, D. B. (1983c). Probabilities of selection and their role for Bayesian modelling in sample surveys. *Journal of the American Statistical Association* **78**, 803–805.

Rubin, D. B. (1983d). Imputing income in the CPS: Comments on "Measures of aggregate labor costs in the United States." In *The Measurement of Labor Cost*. J. Triplett (ed.). University of Chicago Press, pp. 333–343.

Rubin, D. B. (1983e). Discussion of "Statistical record matching for files" by M. Woodbury. In *Incomplete Data in Sample Surveys, Volume 3, Proceedings of the Symposium*. W. G. Madow, and I. Olkin (eds.). New York: Academic Press, pp. 203–212.

Rubin D. B. (1984). Assessing the fit of logistic regressions using the implied discriminant analysis. Discussion of "Graphical methods for assessing logistic regression models" by Landwehr, Pregibone and Smith. *Journal of the American Statistical Association* **79**, 79–80.

Rubin, D. B. (1985). The use of propensity scores in applied Bayesian inference. In *Bayesian Statistics 2*. J. Bernardo, M. DeGroot, D. Lindley, and A. Smith (eds.). Amsterdam: North Holland, pp. 463–472.

Rubin, D. B. (1986). Statistical matching using file concatenation with adjusted weights and multiple imputations. *Journal of Business and Economic Statistics* **4**, 87–94.

Rubin, D. B. (1987). The SIR Algorithm—A discussion of Tanner and Wong's "The Calculation of Posterior Distributions by Data Augumentation." *Journal of the American Statistical Association*. (to appear).

Rubin, D. B., and Schenker, N. (1986). Multiple imputation for interval estimation from simple random samples with ignorable nonresponse. *Journal of the American Statistical Association* **81**, 366–374.

Rubin, D. B., and Thayer, D. T. (1978). Relating tests given to different samples. *Psychometrika* **43**, 3–10.

Sande, I. G. (1983). Hot-deck imputation procedures. In *Incomplete Data in Sample Surveys, Volume 3, Proceedings of the Symposium*. W. G. Madow and I. Olkin (eds.). New York: Academic Press, pp. 334–350.

Santos, R. L. (1981). Effects of imputation on complex statistics. Survey Research Center, University of Michigan Report.

Satterthwaite, F. E. (1946). An approximate distribution of estimates of variance components. *Biometrics Bulletin* **2**, 110–114.

Schenker, N. (1985). Multiple imputation for interval estimation from surveys with ignorable nonresponse. Ph.D. dissertation, Department of Statistics, University of Chicago.

Schenker, N., and Welsh, A. H. (1986). Asymptotic results for multiple imputation. University of Chicago, Department of Statistics Report No. 196.

Scott, A. J. (1977). On the problem of randomization in survey sampling. *Sankhya* **C39**, 1–9.

Sedransk, J., and Titterington, D. M. (1984). Mean imputation and random imputation models. Unpublished technical report.

Sims, C. A. (1972). Comments and rejoinder. *Annals of Economic and Social Measurement* **1**, 343–345, 355–357.

Sudman, S., and Bradburn, N. M. (1974). *Response Effects in Surveys: A Review and Synthesis.* Chicago: Aldine.

Sugden, R. A., and Smith, T. W. F. (1984). Ignorable and informative designs in survey sampling inference. *Biometrika* **71**, 495–506.

Tanner, M. A., and Wong, W. W. (1987). The calculation of posterior distributions by data augmentation. *Journal of the American Statistical Association*. (to appear).

Wachter, K. W., and Trussell. J. (1982). Estimating historical heights. *Journal of the American Statistical Association* **77**, 279–303.

Weld, L. H. (1987). Significance levels from public-use data with multiply-imputed industry codes. Ph.D. dissertation, Department of Statistics, Harvard University.

Wilks, S. S. (1963). *Mathematical Statistics*. New York: Wiley.

Woodbury, M. A. (1983). Statistical record matching for files. In *Incomplete Data in Sample Surveys, Volume 3, Proceedings of the Symposium*. W. G. Madow, and I. Olkin (eds.). New York: Academic Press, pp. 173–181.

Author Index

Afifi, A. A., 159, 244
Althauser, R., 158, 244
Aziz, F., 178, 244

Beale, E. M. L., 191, 244
Beaton, A. E., 191, 244
Berndt, E. K., 225, 244
Binder, D. A., 65, 244
Box, G. E. P., 4, 33, 42, 47, 60, 90, 108, 166, 244
Bradburn, N. M., 17, 244, 250

Cochran, W. G., 4, 10, 13, 57, 69, 148, 158, 193, 221, 232, 233, 243, 244
Cox, D. R., 31, 66, 70, 244

David, M., 227, 244
Dawid, A. P., 36, 72, 245
Dempster, A. P., 108, 132, 189, 191, 197, 245
Diaconis, P., 40, 245
Dijkstra, W., 17, 245
Dillman, D. A., 17, 245

Efron, B., 197, 245
Elashoff, R. M., 159, 244
Ericson, W. A., 42, 44, 47, 65, 74, 245

Feller, W., 40, 245
Fisher, R. A., 79, 93, 108, 148, 245
Ford, B. L., 157, 245

Glynn, R., 234, 236, 237, 238, 245
Greenlees, J. S., 222, 245

Grey, G. B., 70, 248
Groves, G. M., 17, 245

Hajek, J., 58, 74, 245
Hall, B. H., 225, 244
Hall, R. E., 225, 244
Hansen, M. H., 4, 57, 72, 73, 245, 246
Hartley, H. O., 191, 246
Hausman, J. A., 225, 244
Heckman, J., 209, 246
Heitjan, D. F., 17, 246
Herzog, T. N., 125, 133, 151, 179, 181, 182, 183, 185, 246
Hewitt, E., 40, 246
Hinkley, D. V., 66, 70, 244
Holt, D., 70, 246
Huber, P. H., 67, 74, 246
Hurwitz, W. N., 4, 57, 245

James, W., 197, 246

Kahn, R. L., 17, 245
Kilss, B., 178, 244
Kish, L., 4, 57, 246

Laird, N., 108, 132, 189, 191, 197, 234, 236, 237, 238, 245
Lancaster, C., 179, 180, 246
Li, K. H., 99, 101, 102, 109, 116, 139, 192, 194, 200, 246
Lillard, L., 196, 224, 246
Lindley, D. V., 74, 198, 246
Little, R. J. A., 73, 74, 191, 196, 224, 227, 243, 244, 246, 247

Madow, W. G., 4, 57, 58, 72, 73, 196, 245, 246, 247
Marini, M. M., 171, 189, 190, 247
Mitra, S. K., 67, 74, 247
Morris, C. N., 72, 197, 224, 243, 245, 247

Nelson, F. D., 209, 224, 247
Neyman, J., 71, 247
Nisselson, H., 4, 196, 247

Oh, H. L., 70, 179, 247
Okner, B., 187, 247
Olkin, I., 4, 196, 247
Olsen, A. R., 171, 189, 190, 247
Orchard, T., 191, 247

Pathak, D. K., 67, 74, 247
Platek, R., 70, 248
Pratt, J. W., 46, 69, 248

Raghunathan, T. E., 99, 102, 109, 143, 144, 145, 248
Reece, W. S., 222, 245
Rodgers, W. L., 187, 248
Rosenbaum, P. R., 158, 248
Rosenthal, R., 17, 248
Rosnow, R. L., 17, 248
Rubin, D. B., 2, 17, 27, 31, 44, 53, 71, 73, 108, 124, 125, 132, 133, 135, 136, 151, 153, 158, 161, 168, 171, 174, 179, 181, 182, 183, 185, 186, 187, 188, 189, 190, 191, 192, 194, 196, 197, 199, 215, 216, 221, 224, 234, 236, 237, 238, 240, 244, 245, 246, 247, 248, 249

Samuhal, M. E., 227, 244
Sande, I. G., 157, 158, 249
Santos, R. L., 197, 249

Satterthwaite, F. E., 141, 249
Savage, L. J., 40, 246
Schenker, N., 116, 133, 135, 136, 153, 249, 250
Scheuren, F. J., 70, 178, 179, 244, 247
Schluchter, M. D., 191, 247
Scott, A. J., 65, 250
Sedransk, J., 196, 250
Sims, C. A., 187, 250
Smith, A. F. M., 198, 246
Smith, J. P., 196, 224, 246
Smith, T. W. F., 70, 73, 246, 250
Stein, C., 197, 246
Sudman, S., 17, 244
Sugden, R. A., 73, 250

Tanner, M. A., 192, 194, 200, 250
Tepping, J., 72, 73, 246
Thayer, D. T., 199, 249
Tiao, G. C., 4, 33, 42, 47, 60, 90, 108, 166, 244
Titterington, D. M., 196, 250
Triest, R. K., 227, 244
Trussell, J., 243, 250
Tsutakawa, R. K., 197, 245

Vander Zouwen, J., 17, 245

Wachter, K. W., 243, 250
Welch, F., 196, 224, 246
Weld, L. H., 147, 250
Welsh, A. H., 116, 250
Wilks, S. S., 45, 250
Wong, W. W., 192, 194, 200, 250
Woodbury, M. A., 187, 191, 247, 250

Zieschang, K. D., 222, 245

Subject Index

ABB, *see* Approximate Bayesian bootstrap
Actual posterior distribution, definition, 81
Adjusted estimates, need for, 7–11
Adjustment cells, 196
Advantages:
  of imputation, 11–12
  of multiple imputation, 15–16
Advice:
  regarding general analysis methods, 75–81
  regarding general imputation methods, 126–128
Alcohol consumption, *see* Normative Aging Study example
Analyst's model *vs.* imputer's model, 107, 110–112
Approximate Bayesian bootstrap (ABB), 124, 136, 151–152, 232, 240, 242
Approximating posterior distribution of missing data, 192–195
Assuming independent monotone patterns, 190–191
Asymptotic efficiency, relative efficiency, *see* Efficiency, Information
Asymptotic results, standard, 66–67
Asymptotic sampling distributions of procedures, 128–147
Auxiliary matrix of imputations, 2–3, 24
  examples, 182–183, 238
Available-case method, 8–9

Balanced repeated replication (BRR), 197
Bayesian bootstrap (BB), 44, 123–124, 136, 151–152, 242

Bayesian inference, 4, 22–23, 39–74, 55–56, 62–68, 70–112
Bayes's theorem, 33
BB, *see* Bayesian bootstrap (BB)
Behrens-Fisher reference distribution, 90–92
Best-prediction imputation, 13–14, 159, 198
Between-imputation variance, 21
Bivariate data, 172–174, 186–188
Bootstrap, 120–123

Calibration, 62–65, 72–73, 110–112
  absolute, 65
  conditional, 65
  global, 63–65
Census Bureau, 4–6, 9, 12, 157–159, 178, 181, 196, 199
Central Limit theorem, 57–58
Cholesky factorization, 167, 220
Cluster sampling, 31, 36
Coefficient of variation, 218
Cold-deck imputation, 196
Compensatory reading programs, *see* Educational Testing Service (ETS) example
Complete-data and completed-data posterior distributions, 102–107, 110–112
Complete-data methods, desirability of, 7–11
Complete-data posterior distribution, definition, 102
Complete-data statistics, definition, 76

SUBJECT INDEX

Completed-data posterior distribution, definition, 81
Completed-data statistics, definition, 76
Conditional expectations, 33–35
Conditional randomization inference, 70
Congressional seats, 9
Contingency table model, 200
Covariates, X, 28
Coverage of t-based interval estimates:
　large-sample, 114–115
　small-sample, 135–136
CPS–SSA–IRS Exact Match File, 178, 222–229
Current Population Survey (CPS) examples, 5–6, 9, 73, 157, 178, 222–229

Data augmentation algorithm, 192
Data collector's knowledge, 11–12
Data matrix, definition, 2–3
Decennial Census, 9
Decision theory, 72
deFinetti's theorem, 40, 68, 104
Dichotomous outcome, see Logistic regression
Disadvantages:
　of imputation, 12–13
　of multiple imputation, 17–18
Discarding data, 8
　to create monotone pattern, 189–190, 198–199
Discriminant matching, 158
Distinct parameters, 53–54, 174
Distributions:
　Bayesian bootstrap, 44–46
　Behrens–Fisher, 90–92, 108
　beta, 44–46
　chi-squared, 40–42, 60–61
　Dirichlet, 44–46, 124
　F, 60–61
　half normal, 17
　inverted chi-squared, 40–42, 47
　Laplace, 4, 135–136
　lognormal, 71, 135
　normal, 40–42, 46–47, 60–61. See also Normal asymptotic distributions; Normal bivariate data; Normal linear model with hot-deck component; Normal linear regression; Normal model; Normal reference distribution
　t, 40–42, 46–47, 60–61

Wishart, 95, 131
Donee, 196
Donor, 157, 196
Double-coded sample, 6, 10
Double-sampling, 10, 233
Drinking behavior, see Normative Aging Study example

Editing procedures, 2
Educational Testing Service (ETS) example, 4–5, 7–9, 215–222
Efficiency, small sample, 132–133
　asymptotic, 114, 131–132
　see also Information
EM algorithm, 108, 132, 152, 189, 191–192, 194, 199–200
Empirical Bayes, 72, 197–198
Errors in variables model, 152, 198
Estimation task:
　complications with general patterns of missingness, 189–195
　example:
　　with CPS data, 180, 225
　　with ETS data, 219
　　with follow-up response, 235
　　with mixture model, 219
　　with Normative Aging Study data, 235
　　with selection model, 225
　theory, 163–165
　　with monotone-distinct structure and ignorable nonresponse, 175–176
　　with nonignorable nonresponse, 206, 211–215
Exchangeable distributions, 32–33, 40, 68
Excluded values, Y_{exc}, 48
Expanded standard errors, need for, 7–11
Explicit models, 156–157, 173–174
　for general nonmonotone patterns, 191–192
　theory, 160–166
　with univariate Y_i, 166–171
Explicit models vs. implicit models, 181–186
Exposing sensitivity of estimates and standard errors, desirability of, 7–11

Factorization of likelihood, 174
Factorized parameter space, 174
Field efforts to reduce nonresponse, 12
Filematching missingness, 186–188, 198
Filling in data, 8

SUBJECT INDEX

Fixed-response randomization-based coverage, definition, 148
FN, *see* Fully normal imputation method
Follow-up response, 7, 10–11, 229–240
F reference distribution, 77–79
Fully normal imputation method (FN), 123, 136, 152, 242

Gauss–Newton algorithm, 225

Hard-core nonresponse, 230
Highest posterior density regions, 59–60
Hot-deck imputation, 4, 9–10, 22, 25, 120–123, 151, 157–159, 181, 196, 198–199

Ignorable hard-core nonresponse among follow-ups, 233
Ignorable mechanisms, 50–54
Ignorable nonresponse, 22, 154–180
  evidence against, 155
Implicit models, 156–157, 172–173
Importance ratios, 193
Imputation task:
  complications with general pattern of missingness, 189–195
  example:
    with CPS data, 181, 225–226
    with ETS data, 219–220
    with follow-up response, 235–237
    with mixture model, 219–220
    with Normative Aging Study data, 235–237
    with selection model, 225–226
  theory, 161–163
    with monotone-distinct structure and ignorable nonresponse, 177–178
    with nonignorable nonresponse, 206, 211–215
Included values, Y_{inc}, 48
Inclusion in survey, I, 29
Income nonresponse, 5–6, 9, 178–186, 222–229
Increase in posterior variance due to nonresponse, 86
Indexing of units, 27
Indicator variables, I, R, 29–30
Inestimable parameters, 186–188
Inference for nonrespondents, 184–185
Information:
  expected total, 85–86
  fraction missing due to nonresponse, 77, 86, 93–99, 132, 201

missing, 15, 18, 132, 186, 222, 239–240
  and observed, 85–86
  variable fractions missing, 98–99, 141
Internal Revenue Service (IRS), 178
Interval estimates, 21, 25, 54–68, 70, 77
  Bayesian, 49
  large-sample coverages, 134–135
  small-sample coverages, 135–136
Item nonresponse, 5
Iterative algorithm, 192, 225. *See also* EM algorithm

Kernel density estimation, 196

Large sample, *see* Asymptotic results, standard; Asymptotic sampling distributions of procedures; Efficiency, asymptotic
Least-squares:
  computations, 47
  regression, 79–81, 159
Limiting normal distributions, 66–67
Linear interpolation, 153
Linear regression, 69, 79–81, 112, 159, 166–169, 173–174, 181. *See also* Normal linear regression
Logistic regression, 10, 169–170, 180, 197, 224–229, 242
Logit function, 169
Log-linear model, 240
Lognormal model, 111, 136, 148–149
Longitudinal surveys, 171

Marginal distribution, 40
Matching methods, 157–158
Matching variables, 9
Mathematical statistical approaches, 11
Maximum-likelihood estimates, 67
Mean and variance adjusted hot-deck imputation method (MV), 124–125, 136, 151–152, 172, 242
Mean-square random variable, 91
Method of moments, 101, 109
Metric-matching, 158–159
Missing at random, 53–54
Missing-data indicators, 53
Missing information, *see* Information
Missing values, Y_{mis}, 48
Missingness:
  file matching patterns, 186–188, 198
  general patterns, 186–195
  monotone patterns, 170–186, 214–215

SUBJECT INDEX

Mixture modeling, 207–209, 214–222, 234–240
Modeling task:
   complications with general pattern of missingness, 189–195
   example:
      with CPS data, 179–180, 224
      with ETS data, 216–219
      with follow-up response, 235
      with mixture model, 216–219
      with Normative Aging Study data, 235
      with selection model, 224
   theory, 160–161, 163–165
      with monotone-distinct structure and ignorable nonresponse, 171–178
      with nonignorable nonresponse, 205–206, 210–212
Models for data, 39–47
   explicit with univariate Y_i, 166–171
Monotone-distinct structure, 174–186, 214
Monotone missingness, 170–186, 214–215
Multinomial distribution, 124
MV, *see* Mean and variance adjusted hot-deck imputation method

National Academy of Sciences Panel on Incomplete Data, 4
Nested missingness, *see* Monotone missingness
Nominal coverage, 58–59
Nonignorable hard-core nonresponse among follow-ups, 233–234
Nonignorable imputed values from ignorable imputed values, 203–204
Nonignorable nonresponse, 22, 202–243
   with univariate Y_i, 205–210
Noniterative algorithm, 192. *See also* SIR algorithm
Nonnormal reference distributions, 148
Nonscientific surveys, 38
Nonsurvey contexts, 3–4
Normal asymptotic distributions, 66–67
Normal bivariate data, 187–188
Normal linear model with hot-deck component, 168
Normal linear regression, 46–47, 69, 166–169, 173–174, 180, 188–189, 194–195, 216–222, 224–229, 235–239
Normal model, 40–42, 46–47, 56, 68–69, 82–83, 87, 92–93, 135–136

Normal reference distribution, 54, 75
Normality of posterior distributions, 66–67
Normative Aging Study example, 7, 10–11, 234–240
Notation, iv
Not observed values, Y_{nob}, 48

OASDI (old age–survivor and disability income) benefit amounts, 179
Observational studies for causal effects, 158
Observed values, Y_{obs}, 48
Occupation codes, 6–7, 10
Outcome variables, Y, 29

Pairwise-present correlation matrix, 24
Parameter, 27
Partial correlation, 187–188
Patterns of missing values, *see* Missingness
Point estimates, 62
Population fractions of information missing due to nonresponse, 139
Population quantities, 28
Posterior cumulative distribution function, 83
Posterior distribution, 48–50, 62
Posterior mean and variance, 84
Poststratified estimator, 70
Power calculations, 141
P.p.s. (probability proportional to size) estimator and sampling, 36, 69, 193
Predictive mean hot-deck imputation, 168
Prior distribution, 40
Probability distributions, 31–35
Propensity score methods, 158, 197
Proper multiple-imputation methods, 118–128
Public-use data bases, 5–7, 9–10
p-values, *see* Significance levels

Quadratic regression, 112

Random nonresponse, 13
Randomization-based coverage, definition:
   fixed-response, 56–58
   random-response, 58
Randomization-based inference, 4, 22–23, 54–68, 70–74
Randomization-based random-response evaluations, 113–153

SUBJECT INDEX

Randomization-based validity:
  of complete-data inference, 118
  of repeated-imputation inference, 116–117, 119–120
Randomization theory, 39
Rate of convergence, 108, 132
Ratio estimator, 19, 69
Regression model, *see* Linear regression; Logistic regression; Normal linear regression
Relative efficiency of point estimates, large sample, 114
Relative increase in variance due to nonresponse, 77–78
Repeated-imputation, 75, 77
  between variance, 76
  chi-squared statistics, 99–102
  estimator, 76
  inferences, 75–81
  interval estimation, 77
  significance levels:
    based on combined estimates and variances, 77–78
    based on repeated complete-data significance levels, 78–79
  total variance, 76
  within variance, 76
Response behavior, 24
Response in survey, R, 30
Response mechanism, 38–39
  confounded, 39
  ignorable, 51–54
  probability, 39
  unconfounded, 39, 52

Sampling/Importance Resampling algorithm (SIR), 192–195
Sampling mechanism, 35–36
  confounded and nonprobability, 38
  ignorable, 50–54
  probability, 36
  unconfounded, 36, 52
  unconfounded probability, 37
Scalar estimands, derivation of repeated-imputation inferences, 87–94
Scientific sampling techniques, 36
Selection modeling, 207, 209–210, 215, 222–229, 241–242
Sensitivity of inference to models for nonresponse, 7–11, 16–17, 202, 221–222

Sequential sample surveys, 31, 36
Significance levels, 60–61, 77–79
  based on average p-values, 107–108
  based on moments, 137–144
  based on repeated likelihood ratio or χ^2 tests, 99–102
  based on significance levels, 144–147
  derivation of repeated-imputation inferences, 94–102
  randomization-based evaluations, 137–147
Simple random imputation, 14–15, 120–123
Simple random sample, 13, 19, 24–25
Simulating missing values:
  ABB method, 124
  BB method, 123–124
  FN method, 83, 123
  follow-up response, 235–237
  hot-deck, 14–15
  logistic regression, 170
  mixture model, 208, 219–220
  MV method, 124–125
  normal linear model, 167
  normal model, 83, 124
    with hot-deck component, 168
  selection model, 225–226
  SIR algorithm, 193
Simulating posterior mean and variance, 85
Simulation, 96
SIR algorithm, 200
Skewness, 155
Social Security benefits, 178
Stable response, 30–31
Stages of sampling, 31
Statistical matching of files, 186–188
Statistics Canada, 158, 196
Stochastic imputation, 25
Storage requirements, 18
Stratified random sampling, 36
Superpopulation approach, 73
Survey nonresponse, definition, 1–2
Sweep operator, 191, 199
Symmetry of distributions, 17

Transformations, 127–128, 168–169, 203
t reference distribution, 56, 77, 92, 130
Two-stage sampling, 31

Unclassified households, 9
Undercoverage, 243

Unit nonresponse, 5
Univariate Y_i, 156–157, 165–171

Valid inferences, desirability of, 16
Variable-caliper matching, 158
Veterans Administration, 7

Wald test statistics, 71
Waves of follow-ups, 230, 234
Weighted analysis, 151
Weighting cell estimators, 196
Weighting data to handle nonresponse, 8
Within-imputation variance, 21

Applied Probability and Statistics (Continued)

JOHNSON and KOTZ • Distributions in Statistics
  Discrete Distributions
  Continuous Univariate Distributions—1
  Continuous Univariate Distributions—2
  Continuous Multivariate Distributions
JUDGE, HILL, GRIFFITHS, LÜTKEPOHL and LEE • Introduction to the Theory and Practice of Econometrics
JUDGE, GRIFFITHS, HILL, LÜTKEPOHL and LEE • The Theory and Practice of Econometrics, *Second Edition*
KALBFLEISCH and PRENTICE • The Statistical Analysis of Failure Time Data
KISH • Statistical Design for Research
KISH • Survey Sampling
KUH, NEESE, and HOLLINGER • Structural Sensitivity in Econometric Models
KEENEY and RAIFFA • Decisions with Multiple Objectives
LAWLESS • Statistical Models and Methods for Lifetime Data
LEAMER • Specification Searches: Ad Hoc Inference with Nonexperimental Data
LEBART, MORINEAU, and WARWICK • Multivariate Descriptive Statistical Analysis: Correspondence Analysis and Related Techniques for Large Matrices
LINHART and ZUCCHINI • Model Selection
LITTLE and RUBIN • Statistical Analysis with Missing Data
McNEIL • Interactive Data Analysis
MAINDONALD • Statistical Computation
MALLOWS • Design, Data, and Analysis by Some Friends of Cuthbert Daniel
MANN, SCHAFER and SINGPURWALLA • Methods for Statistical Analysis of Reliability and Life Data
MARTZ and WALLER • Bayesian Reliability Analysis
MIKÉ and STANLEY • Statistics in Medical Research: Methods and Issues with Applications in Cancer Research
MILLER • Beyond ANOVA, Basics of Applied Statistics
MILLER • Survival Analysis
MILLER, EFRON, BROWN, and MOSES • Biostatistics Casebook
MONTGOMERY and PECK • Introduction to Linear Regression Analysis
NELSON • Applied Life Data Analysis
OSBORNE • Finite Algorithms in Optimization and Data Analysis
OTNES and ENOCHSON • Applied Time Series Analysis: Volume I, Basic Techniques
OTNES and ENOCHSON • Digital Time Series Analysis
PANKRATZ • Forecasting with Univariate Box-Jenkins Models: Concepts and Cases
PIELOU • Interpretation of Ecological Data: A Primer on Classification and Ordination
PLATEK, RAO, SARNDAL and SINGH • Small Area Statistics: An International Symposium
POLLOCK • The Algebra of Econometrics
PRENTER • Splines and Variational Methods
RAO and MITRA • Generalized Inverse of Matrices and Its Applications
RÉNYI • A Diary on Information Theory
RIPLEY • Spatial Statistics
RIPLEY • Stochastic Simulation
ROSS • Introduction to Probability and Statistics for Engineers and Scientists
RUBIN • Multiple Imputation for Nonresponse in Surveys
RUBINSTEIN • Monte Carlo Optimization, Simulation, and Sensitivity of Queueing Networks
SCHUSS • Theory and Applications of Stochastic Differential Equations

(continued from front)